Nanotechnology Enabled In situ Sensors for Monitoring Health

T0135148

Thomas J. Webster
Editor

Nanotechnology Enabled In situ Sensors for Monitoring Health

 Springer

Editor
Thomas J. Webster
School of Engineering
Brown University
182 Hope Street
Providence, RI 02917
USA
thomas_webster@brown.edu

ISBN 978-1-4899-8244-5   ISBN 978-1-4419-7291-0-(eBook)
DOI 10.1007/978-1-4419-7291-0
Springer New York Dordrecht Heidelberg London

Springer is part of Springer Science+Business Media (www.springer.com)

*This book is dedicated to the best nano
sensors of them all: Mia, Zoe and Ava*

Foreword

Nanotechnology will help the development of new in situ biosensors.

A smart "chip" capable of monitoring, diagnosing, and treating diseases within the human body is a dream for both patients and doctors. Imagine that you have a joint prosthesis implanted with a sensor at the bone-implant interface during surgery. After you come home, the sensor will continuously tell you whether infection is occurring and if it is, will transmit signals to external receivers to tell you that you have an infected implant and will excite an antibiotic reservoir attached to it to treat the infected implant. As a result, you will be monitored and treated without going to the hospital which could significantly delay diagnosis and treatment.

Sounds like a far-far away future technology? Actually this scenario is becoming reality and may be a part of your life in just 10 years, if not earlier. In situ medical sensors, contrary to remote sensors, are those which are placed right at implant sites where they are needed (e.g., the brain, joints, injury sites, etc.) detecting physical, chemical, biological signals, and potentially responding to such stimuli to improve device performance (Fig. 1). Without a doubt, using nanotechnology-derived in situ sensors in the human body means higher accuracy and higher resolution in health monitoring.

There is a long history of using sensors for external monitoring, but the use of internal in situ sensors in the medical community has become promising via recent nanotechnology developments. Because in situ biosensors need to be placed in diverse environments (as small as individual cells), sensor miniaturization is a priority. Advanced micro- and nanofabrication methods make it possible to construct delicate micro/nanodevices that are fully functional as sensors. As just one of many examples discussed in this book, nanotechnology can create self-assembled nanoporous metal membranes with protruding carbon nanotubes that are able to detect cellular level, or even molecular level signals; events clearly important for determining implant success/failure.

How can all this influence the medical device industry? From a medical device industrial perspective, feedback concerning device function is key to their continued success. Such information needs to be prompt, accurate, and efficient. Conventionally, such feedback is either from the patients' subjective description or doctor's postoperative experimental observations (which currently relies on instrumentation

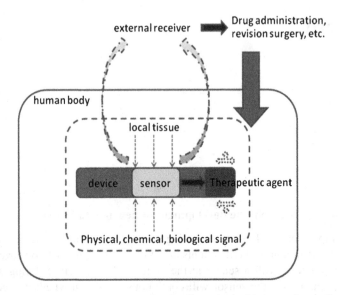

Fig. 1 The internal and external pathway of an intelligent in situ biosensor. Internally, the sensor will activate the combined therapeutic compartment to release drugs that treat implant malfunction. Externally, the sensor will wirelessly send the signal to a receiver so that a doctor can determine the proper treatment for the implant malfunction

not capable of determining cellular events). However, with the development and application of in situ biosensors, feedback can be provided which is no longer restricted by time and method. In situ biosensors can generate new loops of information that can make significant changes to medical device performance and, therefore, to the industry.

In situ biosensors will help us bring new medical devices to the market.

It is common sense that a tremendous amount of preclinical and clinical studies are required by the FDA to prove the safety and efficacy of new medical devices. This process is costly and time-consuming, which may be bad news for patients who are anxiously expecting new solutions. The use of in situ nanotechnology biosensors can change all of this.

Specifically, during the preclinical stage, conventional in vivo experiments obtain information of implant success/failure only after sacrificing animals (with very few exceptions, such as real-time X-rays). If biosensors are implanted within the device, it will make it much easier to obtain continuous information from such animal studies throughout the entire time period of the study. Moreover, measurements from in situ sensors will reflect the body's response more accurately. The same thing will happen in preclinical and clinical studies. The more in situ sensors are used, the more we will learn about patient recovery from implants which will help to develop even better implants. Clearly, however, the most benefit will be less extraction of failed implants from patients through the use of in situ nanotechnology-based sensors.

In summary, the use of biosensors in preclinical and clinical studies will greatly increase the efficiency of preclinical and clinical studies cutting the cost and time the medical device companies spend on these trials. As a result, we, the medical device industry, can provide new technologies and devices to patients in a more timely manner.

In situ biosensors will help us improve current sensor designs and develop the next generation of smart medical devices.

Nanotechnology-enabled in situ sensors could provide unprecedented insights into intact biological environments (from cells to tissues) by providing quantitative, semiquantitative, or qualitative analytical information. Such information will advance our understanding of basic biological interactions between the human body and a medical device (like immune cell reactions, tissue growth, etc.), which in turn will help our researchers develop products in a better manner.

The development of in situ biosensors will also eventually influence our design criteria for future medical devices. Either combined with a medical device or used alone, in situ sensors will be a necessary part of any future smart multifunctional medical system.

Whether you are a medical device engineer, a medical doctor, or a medical device researcher, I would highly recommend this book to you. In this book you will find answers to many interesting questions related to in situ biosensors learning about some already successful stories of their use in different fields. Specifically, you will be able to answer the following questions at the conclusion of this book:

How are nanotechnology, biology, and advanced materials engineering interacting and leading to the development of in situ nanotechnology biosensors?
How are in situ sensors working in different biological environments, e.g. orthopedic and neural applications?
What are the major challenges of current in situ biosensors limiting their clinical use?
Last, but not least, what are the future directions of this field?

The pioneering work on developing and using in situ biosensors is meaningful for all of the medical device community including patients, doctors, medical service providers, and the medical device industry. As this book demonstrates, it is encouraging to see that greater numbers of researchers are devoting their time and energy into this field and it is promising to see that the use of nanotechnology-derived in situ sensors is becoming a large technological success, placing it at a paramount position in medical device history.

West Lafayette, IN   Chang Yao

Preface

Nanotechnology is the use of materials with at least one direction less than 100 nm, so called nanomaterials. Examples of common nanomaterials include particles, fibers, tubes, coating features, etc. all with at least one dimension less than 100 nm. Nanotechnology has already begun to revolutionize numerous science and engineering fields, including, but not limited to fabricating faster and more light-weight computers, stronger buildings (even the consideration of building an elevator from the earth to the moon), improved catalytic devices, and medicine where several nanomaterials have been FDA approved for clinical use. In medicine, numerous researchers are searching for ways to make medicine more personal, and, nanotechnology may provide the answer. Imagine the day when we can utilize sensors placed in various parts throughout the body to determine, in real-time, biological events. Moreover, imagine a day when that same device can send information from inside to outside the body to help a clinician treat a medical problem that would be diagnosed using traditional medical imaging. Lastly, imagine a day when that same device could be programmed to reverse adverse biological events to ensure a healthier, more active patient. This book will examine the role that nanomaterials are playing in the above, specifically, in designing sensors that can diagnosis and treat diseases inside the body.

To introduce the reader to this exciting and fast-moving subject, this book will first provide a forward from the medical device industry concerning the importance of developing in situ sensors to diagnosis and treat diseases in ways we are currently not able to (since, after all, the best way to a fight disease or a medical problem is at the site at which it occurs, not necessarily through conventional systemic drug delivery which takes time and efficacy is lost as drugs pass through the body partially ending up at the disease location). The first chapter will then describe fundamentals of cancer and how in situ sensors are being used to treat cancer. Clearly, cancer remains one of our foremost diseases and nanotechnology-derived sensors are making much progress towards a personalized care of select cancers. The second chapter will then cover the fundamentals of how our tissues heal and how nanosensors can be used to promote tissue healing. Excessive inflammation (often leading to scar tissue growth) and infection often disrupt healing. Thus, the third chapter will cover the fundamentals of inflammation and infection and how sensors are being used to decrease such undesirable events.

The fourth chapter (and those in the remainder of the book), then delves into specific materials used for fabricating sensors. Specifically, the fourth chapter covers how DNA-based materials (due to their unique interactions with proteins, cells, etc.) are being used in nanotechnology-derived sensors. Since DNA has extreme specificity for molecular interactions, it has been widely studied for improving biosensor performance. The fifth chapter then describes unique electrically active materials specifically made for neural applications. The sixth chapter then covers sensors fabricated out of nondegradable and degradable materials exclusively for musculoskeletal applications. Lastly, the book ends with a chapter on the use of carbon nanotubes for biosensor applications. Carbon nanotubes are light weight, strong, and are conductive and, thus, lend themselves nicely to the development of biosensors.

Thus, this book represents a unique combination of engineering, physical sciences, life sciences, and medicine suitable for the design of in situ sensors that can both diagnosis and treat diseases. It covers a wide range of topics from cancer to orthopedics to inflammation and infection. It covers the use of metals, carbon nanotubes, polymers, and DNA in sensors. It provides breadth and depth into this emerging area eventually reaching the dream of developing personalized healthcare through the use of implantable sensors that can both diagnose a disease and treat it on the spot.

Providence, RI   Thomas J. Webster

Contents

Contributors

Yupeng Chen
Department of Orthopaedics, School of Engineering, Brown University, 184 Hope Street, Providence, RI 02917, USA

Batur Ercan
Department of Orthopaedics, School of Engineering, Brown University, Providence, RI 02917, USA

Huinan Liu
Department of Bioengineering, University of California Riverside, 900 University Avenue, Riverside, CA 92521, USA

Justin T. Seil
Laboratory for Nanomedicine Research, Division of Engineering, Brown University, Providence, RI 02917, USA

Sirinrath Sirivisoot
Wake Forest Institude for Regenerative Medicine, Wake Forest University, Health Sciences, 391 Technology Way, Winston-Salem, NC 27157, USA

Nhiem Tran
Department of Orthopaedics, School of Engineering, Brown University, Providence, RI 02917, USA

Phong A. Tran
Orthopedics Research, Rhode Island Hospital, 1 Hoppin Street, Providence, RI 02903, USA

Thomas J. Webster
School of Engineering, Brown University, 182 Hope Street, Providence, RI 02917, USA

Lei Yang
Department of Orthopaedics, School of Engineering, Brown University, Providence, RI 02917, USA

Hongchuan Yu
Department of Chemistry, Boston College, Chestnut Hill, MA 02467, USA

Chapter 1
Nanotechnologies for Cancer Sensing and Treatment

Phong A. Tran

Abstract Current cancer treatment usually involves the application of catheters for anticancer drug delivery to reduce tumor size, followed by surgery to remove the tumor (if possible). Then, more therapy and radiation is used to kill as many tumor cells as possible. The goal of this collective treatment is to target and kill cancerous tissue while affecting as few healthy cells as possible. Due to their nonspecificity, current cancer therapies have poor therapeutic efficacy and can also have severe side effects on normal tissues and cells. Over the last decade, an increase in the survival rate of cancer patients has been achieved, but there is still a need for improvement. In addition, cancer is often diagnosed and treated too late, i.e., when the cancer cells have already invaded and metastasized to other parts of the body. At this stage, treatment methods are limited in their effectiveness. Thus, scientists have been focusing efforts on finding alternative methods to detect cancer at earlier stages and remove such cancerous tissues more effectively. Nanoparticles (i.e., particles with dimensions less than 100 nm in at least one dimension) have become very attractive for improving cancer diagnostics, sensing, and treatment due to their novel optical, magnetic, and structural properties not available in conventional (or micron) particles or bulk solids. Nanoparticles have been extensively studied for various applications including delivering anticancer drugs to tumorous tissues and/or enhancing imaging capabilities to better diagnose and treat cancer. Moreover, nanotechnology has been used to create in situ sensors for cancer cell detection. In this review, recent work related to the improved targeted therapy for specific cancers (whether by developing more specific anticancer agents or by altering delivery methods) are summarized. Discussions on the advantages and disadvantages of the most widely studied nanoparticles used in sensors (i.e., liposome nanoparticles, polymer-based nanoparticles, quantum dots, nanoshells, and superparamagnetic particles) in cancer imaging followed by anticancer drug delivery are highlighted. Lastly, several key advances in the development of sensors to detect cancer are discussed.

P.A. Tran (✉)
Orthopedics Research, Rhode Island Hospital, 1 Hoppin Street, Providence, RI 02903, USA
e-mail: Phong_Tran@brown.edu

T.J. Webster (ed.), *Nanotechnology Enabled In situ Sensors for Monitoring Health*,
DOI 10.1007/978-1-4419-7291-0_1, © Springer Science+Business Media, LLC 2011

Keywords Cancer • Liposomes • Nanoparticles • Quantum dots • Nanoshells • Iron oxide • Sensors

1 Introduction

Nanotechnology is an exciting multidisciplinary field that involves designing and engineering of materials and systems whose structures and components are less than 100 nm in at least one dimension (Ferrari 2005). In this nanometer scale, the properties of objects (such as electrical, optical, mechanical, etc.) significantly differ from those at larger scales, such as the micrometer scale (Klabunde et al. 1996; Wu et al. 1996; Baraton et al. 1997; Siegel and Fougere 1995). Nanotechnology, thus, can be defined as designing, manipulating, and utilizing novel functional structures, devices, and systems on the order of less than 100 nm.

Having the ability to create new products with new characteristics and properties which did not exist before, nanotechnology has shown great potential in a wide range of applications (such as information and communication technology (Heath et al. 1998; Akyildiz et al. 2008), biology and biotechnology (Giaever 2006; Saini et al. 2006), medicine and medical technology (Silva 2004; Jain 2003; West and Halas 2003), etc.). One of the most exciting contributions nanotechnology has made is in medicine and medical research to improve human health. Novel nanometer scale drug delivery systems and nano tools have been developed to aid in disease detection of higher accuracy at earlier stages to simultaneously treat such diseases more effectively (Park 2007, Orive et al. 2005; Kong and Goldschmidt-Clermont 2005; Jain 2003; Shantesh and Nagraj 2006).

Of the tools that nanotechnology has created, nanoparticles have emerged over the last decade as the most promising for disease sensing, treatment, and management, especially for diagnosing and treating cancer. The term cancer nanotechnology refers to the use of nanotechnology to diagnose and treat cancer to increase the survival rate and prolong the lifetime of cancer patients. Nanoparticles have been shown to have unique properties that allow them to deeply penetrate tumors with a high level of specificity. In addition, some nanoparticles also have superior imaging capabilities allowing for a highly sensitive diagnosis of cancer (Shantesh and Nagraj 2006; Brannon-Peppas and Blanchette 2004; Brigger et al. 2002) (see Table 1 for a summary of the advantages and disadvantages of the nanoparticles used in cancer diagnostics and treatment that will be discussed in the chapter). Among many kinds of cancer, bone cancer has been the subject of numerous studies due to its complexity.

It was estimated that 2,380 individuals (1,270 men and 1,110 women) would be diagnosed with bone and joint cancers and 1,470 individuals would die from primary bone and joint cancers in 2008 in the United States (ACS 2008). Primary bone cancer is rare as usually bone cancer is a result of the spread of cancer from

Table 1 Summary of advantages and disadvantages of the use of nanoparticles for cancer treatment to be discussed

Type of nanoparticle	Advantages	Disadvantages	References
Liposomes	Biocompatible; biodegradable; nonimmunogenic; amphiphilic; size, charge, and surface properties of liposomes can be easily changed	Poor control over leakage of drugs; low encapsulation efficacy; poor stability during storage; poor manufacturability at the industrial scale	Torchilin (2005); Tiwari and Amiji (2006); Soppimath et al. (2001); Hans and Lowman (2002)
Quantum dots	Fluorescently bright; large extinction coefficients; high quantum yields; absorption coefficients across a wide spectral range; highly resistant to photobleaching	Composition includes heavy metals which are toxic	Gao et al. (2004a); Dubertret et al. (2002); Ballou et al. (2004); Reiss et al. (2002); Niemeyer (2001); Alivisatos (1996)
Nanoshells	Fine-tunable optical response in a broad region of the spectrum from the near-UV to the midinfrared; can be designed to strongly absorb or strongly scatter light in the NIR region; gold shell is compatible and rigid	Little-known fate following introduction to human bodies	Loo et al. (2004); Loo et al. (2005); Oldenburg et al. (1999); Hirsch et al. (2003)
Superparamagnetic nanoparticles	Controllable nano size; magnetic properties; easy to be directed using an external magnetic field; have controllable, specific Curie temperatures that allow for self-regulated hyperthermia	Hard to be directed to tumors which have a large distance to possible position of magnets; occlusion of blood vessels can occur in the target regions; possible toxic responses to human bodies	Mornet et al. (2004); Wang et al. (2001); Jordan et al. (1996); Wust et al. (2002); Falk and Issels (2001); Kapp et al. (2000); Gerner et al. (2000); Alexiou et al. (2000)
Polymeric nanoparticles	Tailorability of polymer; biocompatibility; biodegradability; various nanoparticle synthesis methods; versatility of drug-loading techniques; controllable drug release characteristics	Some preparation methods use toxic organic solvents; poor drug encapsulation for certain hydrophilic drugs and the possibility of drug leakage	Hans and Lowman (2002); Anderson and Shive (1997); Soppimath et al. (2001)

other organs (such as the lungs, breasts, and the prostate (Miller and Webster 2007)). Because many deaths are officially attributed to the original cancer source, the true numbers of bone cancer-related deaths have been underreported. A common technique to treat bone cancer is the surgical removal of the cancerous tissue followed by insertion of an orthopedic implant to restore patient function. However, it is not always possible to remove all cancerous cells and therefore the remaining tumor cells can redevelop cancer. In these cases, it would be beneficial to have implants specifically designed to prevent the reoccurrence of bone cancer and simultaneously promote healthy bone tissue growth. This same argument can be made for numerous tissues (such as the lung, breast, etc.). This can be achieved by imparting conventional implant materials (such as titanium, stainless steel, ultra-high molecular weight polyethylene, etc.) with anticancer chemistry. To promote healthy bone tissue growth, nanofeature surfacing can be utilized as studies have shown that osteoblast (bone-forming cells) function (from adhesion to proliferation and deposition of calcium-containing minerals) is greater on nanostructured compared to current implant surfaces (which are micron-scale rough and nanoscale smooth) (Webster et al. 1999, 2000a,b, 2001; Webster and Ejiofor 2004; Perla and Webster 2005).

This chapter will first give a brief introduction to the development and characteristics of cancerous tumors and different targeting strategies, then discuss the emerging roles of nanoparticles (such as liposomes, quantum dots (QDs), nanoshells, super-paramagnetic nanoparticles (SPMNPs), and polymeric nanoparticles) in cancer sensing, imaging, and therapeutic purposes.

2 Growth and Other Characteristics of Tumors

2.1 Introduction to Tumors

A tumor starts with a single or a couple of mutated cancerous cells surrounded by healthy, normal tissue. As the cancerous cells replicate, due to its modified DNA, it generally develops at a higher speed than normal cells. The earliest detectable malignant lesions are often referred to as in situ cancers. These are small tumors (a few millimeters in diameter) localized in tissues. At this stage, the tumor is usually avascular, i.e., lacking its own network of blood vessels to supply oxygen and nutrients (Ruddon 1987). Nutrients are provided primarily by diffusion resulting in a slow growth of the tumor. As the cancerous cells develop further, surrounding tissue will not be able to compete for nutrition due to a limited supply of blood. Due to the insufficient nutrient supply, some tumor cells perish, especially those located inside the tumors. Compared to tumor cells at the edge of the tumor, these cells rely solely on diffusion to receive nutrients and to eliminate waste products. The cancerous cells will continue to duplicate and displace surrounding healthy cells until they reach a diffusion-limited maximal size where the rate of tumor proliferation is equal to the rate of tumor cell death. Unless a better connection with the circulatory system is created, tumors cannot grow beyond this diffusion-limited maximal size which is around 2 mm in most tumors (Grossfeld et al. 2002; Jones and Harris 1998). Tumors stay in this stage for years until they initiate the formation of blood vessels (called angiogenesis).

In order to continue growing, tumors need to be vascularized and vascularization of tumors is stimulated by angiogenesis factors (such as vascular endothelial growth factor (VEGF) family, transforming growth factor β, ephrins, cadherin 5-type 2, etc.). The vascularized tumors begin to grow more as nutrition supply is established. Clinically detectable tumors (approximately 10^9 cells, about 1 g mass) may be achieved within a few months or years, depending on cell type. By that time, the tumor will have already gone through approximately two thirds of its lifetime, with about a 30 cell population doubling. If the tumor is unchecked, five more population doublings would produce a cancer mass of about 32 g, five more doublings would give a 1 kg mass tumor of approximately 10^{12} cells. The tumor mass compresses surrounding tissues, invades basement membranes, and metastasizes (Ruddon 1987). This exponential growth of tumors and their metastasis characteristics are the reasons why early cancer detection is of extreme importance.

2.2   Vascularization Process in Tumors (Angiogenesis)

Angiogenesis is the growth of new blood vessels from preexisting blood vessels. Angiogenesis can occur from circulating endothelial precursors, shed from the vessel wall or mobilized from the bone marrow. Angiogenesis starts with pericytes retractring from the abluminal surface of capillaries. After that, endothelial cells release and activate proteases (such as urokinase (uPA), progelatinase A, progelatinase B, etc.) that degrade the extracellular matrix surrounding the existing capillaries. Endothelial cells then migrate, proliferate, and align to form tube-like structures, which ultimately anastomose to form new capillaries (Eatock et al. 2000). Cancer treatment strategies aiming at attacking angiogenesis in tumors can focus on any of those angiogenic stages.

2.3   Characteristics of Tumor Vascular Structures

Tumor vessels are structurally and functionally abnormal. In contrast to normal vessels, tumor vasculature is highly disorganized, e.g., vessels are tortuous and dilated, with uneven diameters, excessive branching, and numerous openings. This may be due to an imbalance of angiogenic regulators, such as VEGF and angiopoietins (Carmeliet and Jain 2000). They may lack functional perivascular cells, which are needed to protect vessels against changes in oxygen or hormonal balance, provide them necessary vasoactive control to accommodate metabolic needs, and induce vascular quiescence. As the result, tumor blood flow is chaotic and variable

(Baish and Jain 2000). Due to the irregular and chaotic structure of the vascular system in tumors, some areas do not obtain enough blood supply and become oxygen starved (hypoxic) (Helmlinger et al. 1997). These areas also become acidic as hypoxic tumors cannot adequately remove waste through the blood stream. The hypoxic tumor cells are difficult to treat if the anticancer drugs are to be delivered through the blood stream. Radiation treatments which create oxygen radicals that attack the DNA of tumor cell are also ineffective to kill these tumor cells.

Tumorous tissues not only have higher vascular density compared to normal tissues but the structure of tumor vessels is also abnormal: the walls of tumor vessels have numerous "openings" (such as endothelial fenestrate, vesicles, and transcellular holes), widened interendothelial junctions, and a discontinuous or absent basement membrane. The endothelial cells are also abnormal in shape, growing on top of each other and projecting into the lumen. Tumor vessels, therefore, are leakier than those of healthy, normal tissues (Carmeliet and Jain 2000). Rapid vascularization also leads to impaired lymphatic drainage systems in tumor tissue. The high vascular density of tumors, their leaky vessels, and defective or suppressed lymphatic drainage in the tumor interstitium comprise the so-called "enhanced permeation and retention (EPR)" effects which anticancer treatment methods can take advantage of (Sledge and Miller 2003; Teicher 2000).

3 Tumor Targeting Methods

3.1 Passive Targeting

Passive targeting is defined as a methodology to increase the target/nontarget ratio of the amounts of delivered drugs primarily by minimizing nonspecific interactions with nontarget organs, tissues, and cells. Nanoparticles are usually taken up by the liver, spleen, and other parts of the reticuloendothelial system (RES) (the macrophage is one of these elements). Depending on surface characteristics, nanoparticles can be taken up by the RES system. Particles with more hydrophobic surfaces (such as poly(propylene oxide), poly(methyl methacrylate), etc.) will be preferably taken up by the liver followed by the spleen and lungs (Brigger et al. 2002). To maximize circulation time and targeting ability, the nanoparticle size should be less than 100 nm (in diameter) and the surface should be hydrophilic to avoid clearance by macrophages. This can be achieved by coating nanoparticles with hydrophilic coatings (such as poly(ethylene glycol) (PEG), poloxamines, or poloxamers (Storm et al. 1995)). Particles that can avoid being taken up by the RES are called "stealth particles." The enhanced permeability and retention (EPR) effect due to leaky vascular structures and impaired drainage system of tumors creates opportunities to increase the transport of drugs from blood vessels into tumor tissues to retain drugs there. Drugs can be designed to have sizes larger than the pore size of blood vessels in healthy tissues (from 2 to 6 nm), but smaller than the pore size of blood vessels in tumor tissues (ranging from 100 to 780 nm) (Yuan et al. 1995; Hobbs et al. 1998).

The drugs will then accumulate preferentially at the tumor site and side effects of the drugs on healthy tissue will therefore be reduced. In addition, due to the ineffectiveness of the lymphatic system's operation at tumor sites, the drugs can be retained in the tumor interstitium longer, therefore increasing treatment effectiveness.

3.2   Active Targeting

Active targeting is the method in which the therapeutic agent is delivered to tumors by attaching it with a ligand that binds to specific receptors that are overexpressed on target cells. Upon binding to the receptors on the target, the particle (now conjugated with ligands) could be internalized into the cell (internalizing ligand) or remain cell-bound without being internalized (noninternalizing ligand). So, what specific receptors are overexpressed on tumor cells? They can be divided into two classes: (1) targets that are preferentially expressed on endothelial cells in tumor blood vessels (e.g., integrin-vβ3 and negatively charged phospholipids) (Hood et al. 2002; Ran et al. 2002) and (2) targets that are overexpressed on tumor cells (e.g., HER2 and disialoganglioside (GD2)) (Park et al. 2002; Pastorino et al. 2003).

The sizes of drug carriers are very critical. One needs the drugs to be transported by blood to the desired site (i.e., tumor) to be taken up by the tumor cells. As previously mentioned, the endothelium of blood vessels in most healthy tissues has a pore size of 2 nm, and 6 nm pores are found in postcapillary venules (Hobbs et al. 1998; Hood et al. 2002). In contrast, the endothelium of tumor vasculature has pore sizes much greater, usually from 100 to 780 nm (Hobbs et al. 1998; Hood et al. 2002). Therefore, as illustrated in Fig. 1, particulate drug carriers of 50–100 nm size can enter the tumor from tumor blood vessels, but cannot get into healthy tissue from healthy blood vessels (Drummond et al. 1999).

Figure 2 gives an example of doxorubicin (DOX)-loaded liposome with phospholipid-anchored folic acid-PEG conjugates (FTL-Dox) accumulated on the side of KB-HiFR tumor cells in vitro (Gabizon et al. 2004). Phospholipid-anchored folic acid is the ligand to target folate receptors (FR) overexpressed on KB-HiFR tumor cells. The DOX fluorescence (orange) is readily recognized in the nucleus sparing the nucleolus.

It is clear that the above choice of nanoparticle size is part of passive targeting when one tries to avoid the uptake of drugs from healthy tissues. It is true that the passive component of drug targeting is important in active targeting systems and should not be overlooked when designing active targeting strategies. It is important because: (1) the majority of a living body comprises nontarget sites. Even the liver, one of the largest drug targets, is only 2% of the weight of the entire body. That is, 98% of the body can be considered to be a nontarget site in this case. As a result, when the drug reaches its target, only a fraction of the applied dose remains after nonspecific capture at nontarget sites. (2) In most cases, before specific

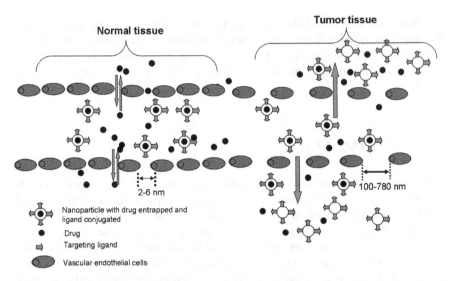

Fig. 1 EPR effects and targeting nanoparticles to the tumor. Nanoparticles with diameters between 50 and 100 nm can avoid entering normal tissue, but can extravasate into tumors and target the tumor cells by ligand–receptor methods and release anticancer drugs to kill the tumor cells

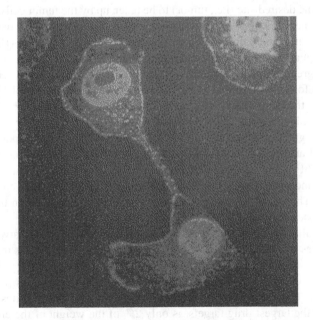

Fig. 2 A confocal fluorescence microscope picture of KB-HiFR tumor cells after 2 h in vitro exposure to 10 μM FTL-Dox (Reprinted with permission from (Gabizon et al. 2004))

ligand–receptor interactions take place in active targeting, the drug–carrier–ligand complex has to be transferred through many other tissues since most targets are located in the extravascular space. Exceptions are cases for intravascular targets such as lymphocytes and vascular endothelial cells. For example, in drug applications via the bloodstream, the drug must first be transferred through the vascular endothelium followed by permeation through the interstitial space to the extravascular targets.

After reaching the tumor site, the drugs will be internalized (i.e., enter tumor cells) with or without internalization of the carrier, i.e., nanoparticles. In the latter, nanoparticles will stay localized in the interstitium surrounding the tumor cells, while the drug, after being released from the drug–nanoparticle conjugate during degradation, will enter tumor cells through diffusion or active transport. The factors that accelerate this degradation to release drugs are possibly the atypical condition of the tumor environment (such as acidic pH and the presence of enzymes and oxidizing agents (Drummond et al. 1999)). Many methods have taken advantage of these factors to design particulates that preferentially disintegrate at acidic pH or increased temperature (Kirpotin et al. 1996). Internalized drugs accumulate in endosomes and then lysosomes. They can only exert pharmacological activities when they exit these organelles and reach the cytosol or the nucleus. If the nanoparticles are also internalized along with the drugs, the internal environment of lysosomes will be the factors that disintegrate the nanoparticles and the drug can then diffuse out of the lysosomes. Between these two strategies of internalizing drugs, the methods that also internalize carrier particles have shown higher delivery efficacy.

4 Liposome Nanoparticles

4.1 Liposomes and their Advantages in Drug Delivery

Liposomes are spherical, self-closed vesicles formed by one or several concentric lipid bilayers with an aqueous phase inside and between the lipid bilayers (Fig. 3a). Therefore, liposomes usually have hydrophilic outer surfaces, hydrophilic inner cores, and hydrophobic matrices in between. Figure 3b gives an example of a transmission electron microscope (TEM) image of liposomes (adapted with permission from reference (Ran et al. 2002)).

There are several advantages of using liposomes for drug delivery in cancer treatment. First, liposomes are highly biocompatible, biodegradable, and nonimmunogenic. These properties are attributed to their composition from naturally occurring lipids. Second, the size, charge, and surface properties of liposomes can be easily changed by adding new ingredients to the lipid mixture before their preparation and/ or by varying the preparation methods. For example, small unilamellar liposomes formed by a single bilayer can be around 100 nm in size, larger unilamellar vesicles have sizes ranging from 200 to 800 nm, and multilamellar vesicles can be as large as 5,000 nm and consist of several concentric bilayers (Torchilin 2005).

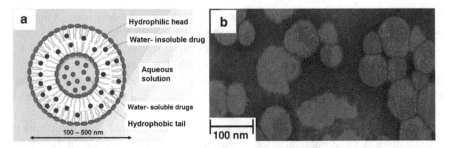

Fig. 3 Schematic structure of a liposome with both water-soluble and water-insoluble encapsulation (**a**) and TEM image of liposome (**b**) (adapted with permission from reference (Ran et al. 2003))

Maybe the biggest advantage of using liposomes for drug delivery is their amphiphilicity (i.e., having both hydrophilic and hydrophobic domains) that allows them to be conjugated with both hydrophilic and hydrophobic therapeutic agents. Water-soluble (hydrophilic) pharmaceutical agents can be entrapped in internal aqueous compartments of liposomes and water-insoluble (hydrophobic) drugs can be placed into the membrane of liposomes (Torchilin 2005; Tiwari and Amiji 2006). Liposomes can also fuse with cell membranes and transfer drugs in the liposome to the inside of the cells. There are two main mechanisms of liposomes releasing encapsulated drugs. One mechanism is based on the development of an affinity reaction. In the other mechanism, the triggered-release mechanism, some elements are incorporated into the liposome and this induces structural changes in the bilayer membrane under an external stimulus, e.g., pH (Hafez et al. 2000; Mizoue et al. 2002; Reddy and Low 2000; Yamada et al. 2005), temperature variation (Anyarambhatla and Needham 1999; Liu and Huang 1994), light irradiation (Benkoski et al. 2006), or ultrasound (Huang and MacDonald 2004). Membrane integrity of liposomes is lost under effects of the external stimulus leading to the release of entrapped compounds (Jesorka and Orwar 2008).

4.2 Passive Targeting Liposomes with PEG Coatings

These traditional particles have several disadvantages as vehicles for drug delivery. They are very quickly captured by the RES (half-life is less than 30 min). They are also instable; therefore, they cannot be used as a delivery agent without modification (Torchilin 2005). The introduction of PEG as a coating on liposomes, as seen in Fig. 4, has prolonged their circulation times in the bloodstream (half-life is about 5 h (Klibanov et al. 1990)) and these vehicles can be targeted to solid tumor sites by the EPR effect (Yokoyama 2005).

PEG is a hydrophilic, nonionic polymer that has been shown to exhibit excellent biocompatibility. PEG molecules can be added to the particles via a number of different routes including covalent bonding, mixing during nanoparticle preparation,

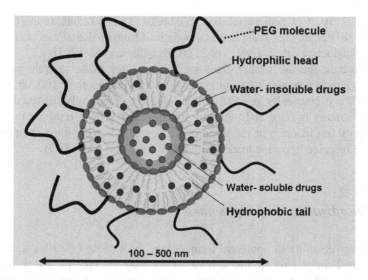

Fig. 4 Schematic illustration of a PEG-coated, drug-loaded liposome nanoparticle

or surface adsorption (Hans and Lowman 2002). Why can PEG coatings prolong liposome circulation time? It has been proposed that there are two main factors that affect the affinity of liposomes to the RES: nonspecific hydrophobic interactions of liposomes with RES cells and a specific opsonization reaction involving some blood component(s) such as immunoglobulin, complement proteins, apolipoproteins, and fetuin (Moghimi and Hunter 2001). PEG-coated liposomes may become more hydrophilic, therefore their nonspecific interactions with RES cells are decreased. Specific interactions with opsonizing proteins are also reduced as PEG, as a coating layer, may act as a shield for liposomes. More interestingly, coating liposomes with PEG can also be designed so that after the PEG-coated liposomes reach the tumor site through the EPR effect, the local conditions of tumor tissues (e.g., acidic pH in tumors) will allow the PEG coating to become detached, promoting intracellular delivery of the drug payload, delivery of oligonucleotides, or delivery of genes (Zalipsky et al. 1999) (Fig. 4).

4.3 Active Targeting with Liposomes

Besides passive targeting of liposomes to the solid tumor site as mentioned above, liposomes can also be conjugated with ligands for active targeting. One of the examples is the incorporation of a ligand of FR to liposomes. FR is significantly overexpressed on many cancer cells compared to their healthy, normal counterparts. Thus, FR expression in malignant tissues can exceed its expression in the corresponding normal tissues by up to two orders of magnitude (Low and Antony 2004). For instance, the folate receptor is overexpressed in ovarian (52 of 56 cases tested), endometrial (10 of 11), colorectal (6 of 27), breast (11 of 53), lung (6 of 18), renal cell (9 of 18) carcinomas, brain metastases derived from epithelial cancers (4 of 5), and neuroendocrine carcinomas (3 of 21) (Garin-Chesa et al. 1993).

To illustrate the use of liposome nanoparticles for folate receptor-mediated cancer targeting, nanometer liposome particles with diameters of 100 nm, loaded with DOX and covalently attached with folate, were effectively delivered to KB xenograft tumors in mice and inhibited greater tumor growth resulting in a 31% higher ($p<0.01$) increase in the lifespan of the tumor-bearing mice compared to those that received liposome loaded with DOX only (Pan et al. 2003).

4.4 Disadvantages of Liposomes

Despite the versatility of liposomal formulations, poor control over leakage of the encapsulated drugs into the blood prior to reaching the specific target, the low encapsulation efficacy, and poor stability during storage and manufacturability at the industrial scale are the main limitations of using liposomes in anticancer applications (Soppimath et al. 2001; Hans and Lowman 2002).

5 Quantum Dots

5.1 Properties of Quantum Dots

Quantum dots (QDs) are nanocrystalline semiconductors composed of an inorganic core (e.g., cadmium, mercury, cadmium selenite, etc.) with a metal shell (e.g., ZnS) that shields the core and renders QDs bioavailable. QDs have diameters ranging from 2 to 10 nm (for core-shell QDs) and 5–20 nm after surface modification (core-shell-conjugate QDs) (Fig. 5a, b). QDs have unique optical and electrical properties (such as their fluorescence emission) and can be tuned from visible to infrared wavelengths depending on their size and composition; especially, they have large absorption coefficients across a wide spectral range and are very bright and photostable enabling their detection for a wide range of applications in vivo (Niemeyer 2001; Alivisatos 1996).

The light-emitting properties of QDs are attributed to quantum effects due to (1) their nanoscale structures and (2) the so-called quantum confinement phenomenon. Quantum confinement is a quantum effect in which the energy levels of a small nanocrystal (smaller than the Bohr exciton radius, about a few nanometers) are quantized, with values directly related to the nanocrystal size (Alivisatos 1996). When exposed to light sources, the cores of QDs absorb incident photons generating electron-hole pairs (characterized by a long lifetime, greater than 10 nanoseconds (Efros and Rosen 2000)). The pair then recombines and emits a less-energetic

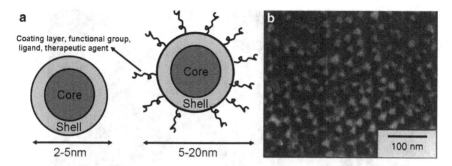

Fig. 5 Schematic quantum dot (QD) structures (**a**) and a TEM image of InAs QDs in a GaAs matrix (**b**) (Adapted with permission from (Grundemann et al. 1995))

photon in a narrow, symmetric energy band (full width at half maximum is typically from 30 to 50 nm) (Michalet et al. 2005). The range of emission wavelength is 400–1,350 nm for QDs with sizes varying from 2 to 9.5 nm not including functional layers (Michalet et al. 2005). In comparison with organic fluorophores, these quantum-confined particles have exceptionally superior properties, therefore offering exciting new opportunities for in vivo imaging, especially in cancer diagnostics and management (Michalet et al. 2005; Larson et al. 2003).

First, fluorescence wavelengths emitted by QDs have higher levels of brightness than those of traditional fluorophores. Single QDs appear 10–20 times brighter than organic dyes (Gao et al. 2004a). QDs have large extinction coefficients that are an order of magnitude larger than those of most dyes (meaning that they can absorb more incoming light per unit concentration of dye) (Dubertret et al. 2002; Ballou et al. 2004) and high quantum yields (close to 90% (Reiss et al. 2002)) (meaning a high amount of light emitted over that absorbed) (Bailey and Nie 2003).

Second, QDs have large absorption coefficients across a wide spectral range (Niemeyer 2001; Alivisatos 1996). This property allows one to simultaneously excite multiple QDs of different emissions with a single excitation wavelength. Their emission spectra have very distinct and narrow wavelengths, which allow independent labeling and identification of numerous biological targets (Han et al. 2001; Gao and Nie 2003, 2004b). This is very useful in studying tumor pathophysiology, which requires one to be able to distinguish and monitor each component of the tumor microenvironment under dynamic conditions. For example, researchers have used two-photon microscopy to image blood vessels within the microenviroment of a tumor using PEG-coated QDs. As seen in Fig. 6, good contrast between cells, matrix, and the leaky vascular was evident.

This suggests the use of QDs' fluorescence contrast imaging for noninvasive diagnostics of human tumors.

Another advantage of QDs is that they are highly resistant to photobleaching than their organic counterparts. This feature is of great importance for three-dimensional (3D) imaging where there is a constant bleaching of fluorophores during acquisition of successive layer-by-layer scanning, which compromises the

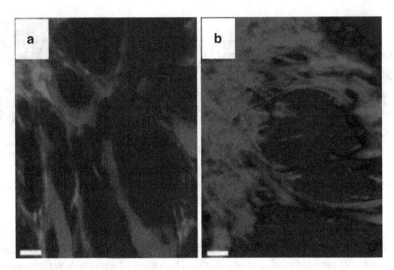

Fig. 6 (**a**) Concurrent imaging of both QDs with a 470 nm emission maximum and green fluorescent protein (GFP) provides a clear separation of the vessel from GFP-expressing perivascular cells and (**b**) vessels (QDs with a 660 nm emission maximum) micelle preparation were imaged simultaneously with the second harmonic generation signal; the image represents a projection of a stack of 20 images at an interval of 2 μm per slice. *Scale bars* represent 50 μm (Reprinted with permission from reference (Stroh et al. 2005))

correct reconstruction of 3D structures. For instance, it was demonstrated that, after 80 min of constant illumination, signal intensity from QDs' fluorescently labeled Xenopus embryos stayed unchanged while dextran-labeled controls were completely photobleached (Dubertret et al. 2002).

5.2   Quantum Dots in Cancer Imaging and Treatment

5.2.1   Active and Passive Targeting for QDs

Having superior light-emitting properties, QDs are excellent candidates for tumor imaging if they can be effectively delivered to tumor sites. Both passive and active targeting mechanisms are being aggressively pursued in order to achieve this goal. Numerous studies have reported methods to lengthen circulation time of QDs in the blood and target QDs to cancerous tissues (Gao et al. 2004a; Åkerman et al. 2002; Rosenthal et al. 2002). Coating QDs with polymers (such as PEG) to avoid uptake by the RES, thereby improving circulation time, is an attractive approach being actively studied. For example, PEG-coated CdSe-ZnS (core-shell) QDs have been shown to circulate longer in the mouse blood stream (half-life more than 3 h) compared to small organic dyes, which are eliminated from circulation within minutes after injection (Ballou et al. 2004). These PEG-coated QDs were also demonstrated to fluorescence after at least 4 months in vivo. It is believed that these PEG-coated QDs are in an intermediate size range in which they are small and hydrophilic enough to slow down opsonization (that is, the alteration of a particle's surface either by the attachment of complement proteins or antibodies specific for the antigen, so that the particle can be ingested (phagocytosed) by phagocytes, macrophages, and/or neutrophils) and reticuloendothelial uptake, but they are large enough to avoid renal filtration.

Taking this a step further, some researchers have conjugated PEG-coated QDs with ligands that can recognize cell membrane receptors (such as HER2, disialoganglioside (GD2)) that are overexpressed on cancer cells. This approach is being pursued both for active and passive targeting to achieve the goal of maximizing accumulation of QDs at tumor sites. For example, a new class of QD conjugates containing an amphiphilic triblock copolymer for in vivo protection, targeting ligands for tumor antigen recognition, and multiple PEG molecules for improved biocompatibility and circulation has been developed (Gao et al. 2004a). In this study, CdSe-ZnS (Core-Shell) QDs encapsulated in a copolymer layer, tri-n-octylphosphine oxide (TOPO), coated with PEG and conjugated to a prostate-specific membrane antigen monoclonal antibody (PSMA) were prepared. PSMA has previously been identified as a cell surface marker for both prostate epithelial cells and neovascular endothelial cells (Chang et al. 2001) and has been selected as an attractive target for both imaging and therapeutic intervention of prostate cancer. These QD conjugates have been targeted to tumor sites in mice through both active and passive targeting mechanisms and have enhanced fluorescent imaging ability.

5.2.2 QDs in Drug Delivery and Therapy for Cancers

In addition to utilizing QDs in imaging, there is a growing interest in using QDs for drug delivery and therapy. Firm control over the size of QDs during nanocrystal synthesis will allow for the evaluation of the nanoscale size effect on the delivery efficiency and specificity, and therefore, identification of the optimal dimensions of drug carriers. High surface-to-volume ratios of QDs allow scientists to impart multiple functionalities on single QDs to develop a multifunctional vehicle while keeping the overall size within the optimal range. For example, by coating a QD with an amphiphilic polymer layer (such as poly(maleic anhydride-alt-1-octadecene) (PMAO), octylamine-modified poly(acrylic acid), etc.), hydrophilic therapeutic agents and targeting biomolecules (such as antibodies, peptides, and aptamers) can be immobilized onto the hydrophilic side of the amphiphilic polymer and small molecular hydrophobic drugs can be embedded between the inorganic core and the amphiphilic polymer coating layer as illustrated in Fig. 7a. Figure 7b shows an image of fixed breast cancer SK-BR-3 cells incubated with a monoclonal anti-Her2 antibody (which was used to bind to the external domain of Her2) and a goat anti-mouse IgG conjugated to QDs with an emission maximum at 630 nm (QD 630-IgG). Her2 was clearly labeled with QD 630-IgG (the outer "ring" in the picture).

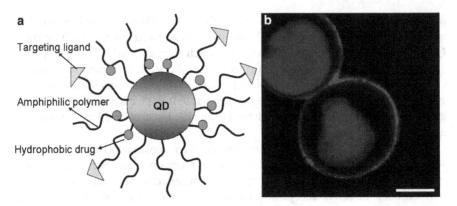

Fig. 7 (**a**) Schematic of QDs fabricated to carry drugs and a targeting ligand and (**b**) fixed breast cancer SK-BR-3 cells incubated with a monoclonal anti-Her2 antibody and a goat anti-mouse IgG conjugated to QDs with an emission maximum at 630 nm (QD 630-IgG). Her2 was clearly labeled with QD 630–IgG (shown in the outer ring). The nuclei were counterstained with Hoechst 33342 (the inside region). *Scale bar* represents 10 μm (Adapted with permission from (Wu et al. 2002))

This integrated nanoparticle may serve as a magic bullet that will not only identify, bind to, and destroy tumor cells but will also emit detectable signals for real-time monitoring of its trajectory.

5.3   Disadvantages of QDs

The main disadvantages of QDs are due to their composition of heavy metals (e.g., Cd, Pb, and Se which are known to be toxic to vertebrate systems at parts-per-million concentrations (Hardman 2006)) and the instability of uncoated QDs when exposed to UV radiation which leads to the release of heavy metal ions. For example, the free cadmium in solution can bind to sulfhydryl groups of critical mitochondrial proteins. Thiol group inactivation then leads to oxidative stress and mitochondrial dysfunction (Rikans and Yamano 2000). The instability of QDs under UV exposure is due to the fact that the energy of UV radiation is close to that of covalent chemical bonds, therefore UV exposure can dissolve the semiconductor particles in a process known as photolysis, which releases toxic ions such as cadmium ions (Nie et al. 2007). For example, under oxidized (30 min exposure to air) or long UV radiation (2–8 h) conditions, even QD concentrations of 0.0625 mg/mL were found to be highly toxic (Derfus et al. 2004). The mechanisms of oxidation of Cd-based QDs have been suggested to occur via either TOPO-mediated or UV-catalyzed surface oxidation resulting in the removal of surface atoms and release of Cd^{2+} ions (Derfus et al. 2004).

Generally, QD toxicity depends on the effects of multiple factors from both individual QD physicochemical properties and environmental conditions. It has been shown that QD size, charge, concentration, outer coating bioactivity (i.e., capping material, functional groups, etc.), and oxidative, photolytic as well as mechanical stability are factors that determine QD toxicity (Hardman 2006). For example, investigators have tested the toxicity of CdSe QDs in liver cultures and found that it was dependent on processing parameters during synthesis, exposure to ultraviolet (UV) light, and surface coatings (Derfus et al. 2004). The CdSe (core only, no shell) QDs were found to be noncytotoxic under standard conditions of synthesis and water-solubilization with mercaptoacetic acid (MAA). However, TOPO-coated QDs, which were initially subjected to air for 30 min and then modified with MAA, showed a dramatic dose-dependent decrease in primary hepatocyte viability (from 98 to 21% at a QD concentration of 62.5 µg/mL). These researchers also capped CdSe QDs with 1–2 monolayers of ZnS to study the effect of capping on cytotoxicity. They found that the CdSe core was intact after 12 h of oxidation in air. However, high levels of free Cd in solution (~40 ppm) after 8 h of UV photooxidation (power density of 15 mW/cm^2 preceded by 12 h of oxidation in air) were observed. This finding indicates that ZnS capping was effective in eliminating cytotoxicity due to oxidation by air, but did not fully eliminate cytotoxicity induced by UV photooxidation.

To overcome those cytotoxicity problems, QDs have been coated with PEG or encapsulated in micelles (i.e., vesicles formed by aggregation of surfactant molecules dispersed in a liquid colloid), which can limit the release of toxic metals in response to UV light exposure (Gao et al. 2004, 2005; Dubertret et al. 2002; Stroh et al. 2005). More work is still needed to develop optimal coatings and, thus, encapsulation that prevents the release of heavy ions from QDs.

6   Nanoshells

6.1   Structure of Nanoshells

Metal nanoshells are spherical nanoparticles consisting of a dielectric core such as silica covered with a thin layer of metal (typically gold). Gold nanoshells have optical properties similar to colloidal gold nanoparticles (such as strong optical absorption due to the collective electronic response of metal to light). However, gold nanoshells show a strong dependence of optical response due to the relative size of the core and the shell's thickness. By varying the core size and the thickness of the gold layer, the color of these nanoparticles can be fine-tuned across a wide spectrum spanning the visible and near-infrared regions (NIRs). Therefore, nanoshells show a great promise in biomedical imaging and therapeutic applications in general and cancer imaging and treatment in particular (Loo et al. 2004, 2005; Hirsch et al. 2003; Chen and Scott 2001). In addition, as with other types of nanoparticles, especially gold nanoparticles, drugs can be conjugated to nanoshells for drug delivery to tumors as seen in Fig. 8.

The optical properties of nanoshells are attributed to the plasmon resonance at the dielectric–gold interface. Plasmon resonance is the phenomenon in which

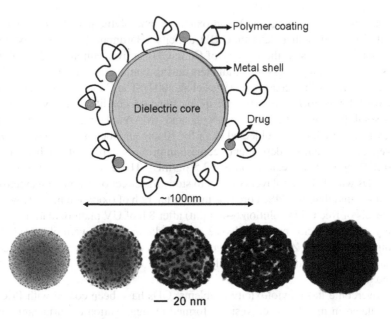

Fig. 8 (*Top*) Schematic illustration of a nanoshell. Polymers (such as hydrogels) can be coated onto a nanoshell surface and drug molecules can be entrapped within the polymer matrix. (*Bottom*) Transmission electron microscope images of gold/silica nanoshells during shell growth (Reprinted with permission from reference (Loo et al. 2004)). From left to right are the nanoshells with a growing metal shell

light induces collective oscillations of conductive metal electrons at the dielectric–gold interface (Hirsch et al. 2003). The absorbing and scattering properties of the particle will then depend on the particle's plasmon resonance. Although many bulk metals have plasmon resonance behavior, it has been observed over a very small region of the visible spectrum. Depending on the relative thickness of the core and shell layers of a nanoshell, its plasmon resonance and the resultant optical absorption can be tuned across a broad region of the spectrum from the near-UV to the midinfrared. This range spans the NIR (Oldenburg et al. 1999), a region where optical absorption in tissues is minimal and penetration is optimal (Weissleder 2001).

The sensitive dependence of optical properties on the core diameter-shell thickness ratio can be understood by the hybridization model of Prodan et al. (2003) in which plasmon resonance of a nanoshell is considered as the result of the interaction between the nanosphere plasmon and cavity plasmon which are electromagnetic excitations that induce surface charges at the outer and inner interfaces of the metal shell, respectively. The sphere plasmon itself depends on the diameter of the sphere and the cavity plasmon is a sensitive function of the inner and outer radius of the metallic shell (Aden and Kerker 1951). In addition, the interaction of the sphere and cavity plasmons depends on the thickness of the shell layer.

6.2 Optical Properties of Gold Nanoshells

As previously stated, the optical response of nanoshells depends drastically on the relative size of the dielectric core and thickness of the gold layer. The ability to fine-tune nanoshells to have optical resonance varied over a broad region ranging from the near-UV to the mid-infrared is very important for tissue imaging. It has been demonstrated that nanoshells can be developed to highly scatter light within this spectral region as would be desired for many imaging applications; alternatively, nanoshells may be engineered to function as highly effective NIR absorbers permitting photothermal-based therapy applications as well (Loo et al. 2004); dual scattering/absorbing nanoshells can also be fabricated (Loo et al. 2005). For instance, the conventional NIR dye indocyanine green has an absorption cross-section of 10^{-20} m^2 at ~800 nm, while the cross-section of absorbing nanoshells can be as high as 4×10^{-14} m^2, approximately a million-fold increase in the absorption cross-section (Hirsch et al. 2003). This indicates that those nanoshells are over a million-fold more (compared to the conventional dye) likely to encounter an absorbing event and convert that light into thermal energy, which can be used to kill cancer cells or improve chemotherapy and radiotherapy efficiency.

6.3 Nanoshells in Cancer Diagnostics and Treatment

Gold nanoshells, having attractive optical properties as just discussed, offer promises for biomedical sensing and therapeutic applications, especially in cancer diagnostic and treatment. In addition to attractive optical properties, nanoshells also have rigid structures due to metallic shells, therefore offering improved stability. Therapeutic agents and other targeting moieties can be readily conjugated to the surface of nanoshells to specifically target and destroy cancer cells (Loo et al. 2004). To enhance biocompatibility and improve blood circulation, nanoshells are usually coated with stealthing polymers (such as PEG), even though the gold surfaces of nanoshells are generally considered to be biocompatible (Tang et al. 1998).

Researchers have developed a novel approach to combine cancer diagnostics and therapeutics based on the use of gold nanoshells as near-infrared (NIR) absorbers. In one exciting in vitro study (Loo et al. 2004), 120 nm diameter silica core-10 nm gold shell nanoparticles were fabricated to provide peak optical scattering and absorption efficiencies in the NIR region (~800 nm). These nanoshells were used for combined imaging and therapy of SKBr3 breast cancer cells. To target these cancer cells, anti-HER2 was conjugated onto the nanoshells to specifically deliver the conjugated anti-HER2-nanoshells to SKBr3 breast adenocarcinoma cells which overexpressed HER2. For cells incubated with conjugated anti-HER2 nanoshells, significantly increased scatter-based optical contrast due to nanoshell binding was achieved as compared to the control cell groups. The second function of these conjugated nanoshells, i.e., therapy, was achieved via photothermal therapy. Incubated cells were exposed to NIR irradiation (820 nm, 0.008 W/m² for 7 min). Cell death was observed only in the SKBr3 breast cancer cells incubated with anti-HER2 nanoshells and then treated with a NIR laser. Cell death was not observed in cells treated with either nanoshells conjugated to a nonspecific antibody or cells exposed to NIR light alone. Improved nanoshell binding to cell surfaces overexpressing HER2 was evident from greater silver staining intensity in cells exposed to anti-HER2 nanoshells compared to controls.

In vivo studies have also shown promising initial results for the use of gold nanoshells for cancer detection and treatment. In one study (Hirsch et al. 2003), investigators fabricated nanoshells of 110 ± 11-nm diameter core and a 10 nm thick gold shell. These gold nanoshells absorb light in the NIR range having a peak absorbance at 820 nm. A monolayer of PEG was assembled onto the gold nanoshell surfaces to achieve higher stability. These nanoshells were injected interstitially (~5 mm) into a solid tumor volume in immunodeficient (SCID) mice. Within 30 min of injection, tumor sites were exposed to NIR light (820 nm, 4 W/cm², 5 mm spot diameter, <6 min).

It was demonstrated in that study that tumors reached temperatures which caused irreversible tumor damage ($\Delta T = 37.4 \pm 6.6°C$) within 4–6 min. In contrast, controls (which were exposed to a saline injection rather than nanoshells) experienced significantly reduced average temperature elevations after exposure to the same NIR light levels ($\Delta T < 10°C$). Gross pathology indications of tissue damage showed coagulation, cell shrinkage, and loss of nuclear staining in nanoshell-treated tumors; no such changes were found in control tissues (Hirsch et al. 2003).

6.4 Disadvantages of Nanoshells

Although nanoshells with attractive optical properties (which can be systematically fine-tuned) offer great promises in cancer detection and treatment, limitations still exist. The biggest disadvantage of nanoshells is probably their little-known fate following introduction into human bodies. These particles have a very small size which might present some potential undesired effects on the organs and tissues in our bodies. More intensive in vivo animal studies are currently being completed in order to investigate both applications and limitations of nanoshells in cancer imaging and treatment.

7 Superparamagnetic Nanoparticles (SPMNPs)

SPMNPs have been extensively studied in biomedical applications and, in particular, cancer diagnostics and treatment (Mornet et al. 2004; Wang et al. 2001; Jordan et al. 1996; Wu et al. 2002; Falk and Issels 2001; Kapp et al. 2000; Gerner et al. 2000;

Alexiou et al. 2000). They have some attractive properties that make them suitable for tumor imaging and tumor treatment. First, they have a controllable nano size. Second, they possess magnetic properties that render them unique in magnetic-based imaging techniques and magnetic-based treatment methods. For example, SPMNPs can be conjugated with drugs, targeted to tumors using magnets, then they can absorb energy from a time-varying magnetic field to be heated resulting in the release of drugs while heat elevation can cause chemotherapy and radiation therapy to be more effective or can directly destroy tumor tissues. In this section, we review the use of SPMNPs in magnetic resonance imaging (MRI), hyperthermia treatment, and magnetic-based drug delivery.

7.1 SPMNPs Used as Magnetic Contrast Agents in MRI

MRI works by the dual application of an external magnetic field B_o (up to 2 T) and a transverse (i.e., transverse to B_o) radio frequency (RF) pulse on protons which are present in large amounts in biological tissues (e.g., in water molecules, membrane lipids, proteins, etc.). Each proton has a tiny magnetic moment that can be aligned by applying an external magnetic field. The external magnetic field first applies and aligns a magnetic moment of each proton. The RF pulse is then directed to the region of interest, and the protons absorb energy from the pulse and spin in a direction different than the direction of the external field. The RF pulse then will be turned off, the protons now begin to return to their original alignment with the external magnetic field and release excess stored energy through a process called relaxation. When this happens, a signal will be given off and the MRI system can detect this signal and process it to transform it into images of the tissue area of interest.

Relaxation can be divided into two independent processes: (1) longitudinal relaxation, in which the longitudinal component of the proton's magnetic moment returns to be aligned with an external field B_o and this is termed T_1-recovery; (2) transverse relaxation, in which the transverse (i.e., transverse to B_o) component of the proton's magnetic moment vanishes and this is called T_2-recovery. Different tissues have different T_1 and T_2 values, therefore they can be differentiated by MRI scanning. Both T_1 and T_2 can be shortened by the use of magnetic contrast agents. Magnetic contrast agents work by altering local magnetic moments in the region of interest, therefore relaxation signals can be enhanced. For example, T_2 can be shortened according to (1):

$$1/T_2* = 1/T_2 + \gamma \Delta B_o / 2. \tag{1}$$

Where ΔB_o is the variation in the field brought about by variations in the local magnetic field inhomogeneity or local magnetic susceptibility of the system due to the presence of magnetic enhance agents (Leach 1988; Brown and Semelka 1999). Figure 9 shows an example of TEM images of Fe_3O_4 SPMNPs and schematically illustrates how SPMNPs can affect T_2 recovery time.

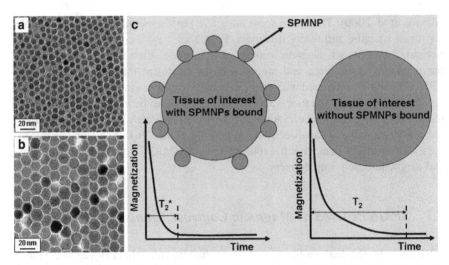

Fig. 9 TEM bright field images of (**a**) 6 nm and (**b**) 12 nm Fe_3O_4 SPMNPs deposited from their hexane dispersion on an amorphous carbon-coated copper grid and dried at room temperature (Adapted with permission from reference (Sun et al. 2004)), and (**c**) schematic illustration of SPMNPs bound to a tissue of interest, thereby shortening T_2 recovery time

Compared to traditional MR contrast agents (such as gadolinium), SPMNPs have several advantages. First, SPMNPs are in the size range in which they can be delivered to the desired tissues more effectively. They are often composed of a superparamagnetic core composed of iron oxides such as magnetite Fe_3O_4, maghemite (γ-Fe_2O_3), or other insoluble ferrites (generally from 3 to 10 nm in diameter) with or without polymer coatings such as dextran (Mornet et al. 2004). In particular, in the case of MRI for tumor imaging, SPMNPs can be delivered to tumors using passive targeting methods. Second, they have much greater magnetic susceptibility (specifically, two or three orders of magnitude greater than traditional gadolinium superparamagnetic particles (Wang et al. 2001)), meaning they are more responsive to an external magnetic field, than traditional MR contrast agents. For example, superparamagnetic iron oxides are effective as T_2 relaxation enhancers due to the induction of local magnetic field inhomogeneity in the surrounding tissues leading to a stronger shortening of the recovery time as in (1).

7.2 SPMNPs in Hyperthermia Treatment for Cancer

Hyperthermia is when temperatures of certain organs or tissues are raised to 41–46°C. If the temperature is raised to above 47°C, tissue destruction occurs and this process is called thermoablation (Jordan et al. 1999). Thermoablation is characterized by acute necrosis, coagulation, or carbonization of tissue. It is clearly undesirable in a clinical situation due to systemic side effects and further clinical complications (Moroz et al. 2002). Modern hyperthermia trials focus mainly on the optimization of thermal homogeneity at moderate temperatures (42–43°C) in the target volume. SPMNPs have the potential to solve this optimizing problem.

As it is well known, tumor cells are more susceptible to heat increases than healthy, normal cells (Chen et al. 2007). Clinical experiments have been taking advantage of the higher sensitivity of tumor cells to temperatures in the range of 42–45°C than normal tissue cells (Overgaard and Overgaard 1972; Moroz et al. 2001). Hyperthermia treatment for cancer works by raising the tumor temperature resulting in damage in the plasma membrane, the cytoskeleton, and the cell nucleus (Shellman et al. 2008). In addition, certain regulatory proteins, cytokines, or kinases are influenced by hyperthermia, which leads to changes in the cell cycle and can even induce apoptosis (i.e., cell death driven by the cell regulatory system itself) (Fairbairn et al. 1995; Sellins and Cohen 1991). Cancer cells are more vulnerable to elevated temperatures compared to normal, healthy cells because (Chen et al. 2007):

1. Normal cells reside close to normal blood streams which can dissipate heat more effectively compared to cancerous cells which have abnormal blood flow and
2. Cancerous cells have a more acidic surrounding environment, therefore they are more susceptible to hyperthermia

There are several different methods for creating local heat elevation (such as using microwave radiation, capacitive or inductive coupling of radiofrequency fields) by implanting electrodes, by ultrasound, or by lasers. Magnetic hyperthermia utilizes losses during the magnetization reversal process of the particles when exposed to alternating magnetic fields to convert such losses to heat. In magnetic hyperthermia, superparamagnetic particles are directed and concentrated at the desired location. An alternating magnetic field will be applied and this application results in the magnetization reversal process of the particles. Energy lost in this process is converted to heat.

The advantages of using SPMNPs in hyperthermia are as followed. First, they can be effectively directed to tumor tissues using magnetic fields and hence can avoid heating up normal tissues. For example, in one study, superparamagnetic magnetite (Fe_2O_3/Fe_3O_4) particles (core size 3.1 ± 0.7 nm) coated with modified dextran were targeted to C3H mammary carcinoma in the right hind leg of mice using magnets after intratumoral injection of the particles. With an applied 50 mT magnetic field, the iron concentration at the tumor site was shown to increase 2.5-fold compared to the control in which no magnets were used for targeting (Jordan et al. 1996).

Second, SPMNPs can be readily conjugated with therapeutic agents and delivered to tumors. Upon reaching the tumor site, an alternating magnetic field will be applied causing the particles to be heated, melting the thermosensitive surface, leaking the drug at the tumor site and, at the same time, heating the tumor. This strategy has a dual effect on cancer treatment as researchers have shown increased chemotherapy efficacy when combined with hyperthermia (Wust et al. 2002; Falk and Issels 2001; Kapp et al. 2000; Gerner et al. 2000).

Another advantage of SPMNPs in hyperthermia is that they possess an appropriate and specific Curie temperature (T_c) (i.e., the temperature above which SPMNPs lose their magnetic properties and, in turn, their coupling with the exter-nal magnetic field) that limits the hyperthermia at the predetermined temperature. For example, copper nickel (Cu/Ni) alloy magnetic nanoparticles were synthesized and coated with PEG and these particles have a Curie temperature in the range of 43–46°C (Chatterjee et al. 2005). This low range of Curie temperature of the Cu/Ni magnetic nanoparticles is very desirable in hyperthermia applications. Once the temperature of the magnetic nanoparticles increases to above T_c (from 43 to 46°C in this case), the external magnetic field will stop interacting with the nanoparticles; hence, further tissue heating will not occur.

7.3   Magnetic Targeting of SPMNP: Drug Conjugates

A clear advantage of using SPMNPs is that they can be directed to tumor sites using an external magnetic field. Using this method, SPMNPs conjugated with anticancer drugs can be delivered to tumors locally to release therapeutic agents and the con-jugate can also be held by the magnetic field in a place at the desired site leading to increased drug delivery efficacy. For example, investigators have used 100 nm (hydrodynamic diameter) iron oxide superparamagnetic particles bound to mitox-antrone (that is an agent used in the treatment of certain types of cancer including metastatic breast cancer and acute myeloid leukemia) to target squamous cell car-cinoma in rabbits (Alexiou et al. 2000). A magnetic flux density of a maximum of 1.7 T was used to produce an inhomogeneous magnetic field that was focused on the tumor in order to attract SPMNPs. A very high concentration of SPMNPs within the tumor after intra-arterial infusion was achieved. Importantly, this appli-cation method resulted in a complete and permanent remission of the squamous cell carcinoma compared with the control group (no treatment) with no signs of toxicity.

In another example, investigators used a ferrofluid (i.e., colloidal solutions of iron oxide SPMNPs surrounded by anhydroglucose polymers) to which the drug epiru-bicin, a well-known antibiotic antracylin that has been used for the treatment of solid tumors (Bonadonna et al. 1993), was chemically bound (Lubbe et al. 1996a). This conjugating method allowed a reversible ionic binding of the drug such that certain physiological parameters of tumors (such as osmolality, pH, temperature) affected desorption of the drug from the particles in the tumor tissue. Their animal experi-ments (Lubbe et al. 1996a) showed for the first time that documented tolerance and efficacy was observed in mice and rats, in which no LD50 (i.e., a dose at which 50% of subjects will die) could be found for the ferrofluid. This success led to clinical trials in humans. The ferrofluid was directed and concentrated in the desired sites using permanent magnets that generated a magnetic field of 0.8 T in tumors located near the body surface (Lubbe et al. 1996b). It was shown that the ferrofluid could be successfully directed to the tumors in about half of the patients.

7.4 Disadvantages of SPMNPs

In spite of the very exciting properties SPMNPs can offer, there are still several problems associated with using these nanoparticles in cancer diagnostics and therapeutics, especially for magnetically targeted drug delivery (Lubbe et al. 1999). The first limitation lies in the use of magnets for attracting superparamagnetic particles. When dealing with the tumors which have a large distance to possibly position the magnets, magnetic field strength at the tumor site will not be enough to attract the SPMNPs. Secondly, occlusion of blood vessels can occur in the target regions as the result of the high accumulation of magnetic carriers. In addition, toxic responses of the body to the SPMNPs can also hinder the application of such particles. All of these potential disadvantages need to be carefully considered.

8 Polymeric Nanoparticles

Polymers have become attractive in formulating drug-loaded nanoparticles for treating cancers due to the tailorability of polymers, biocompatibility and biodegradability properties, various nanoparticle synthesis methods, versatility of drug-loading techniques, and controllable drug release characteristics. In this section, common fabrication methods will be briefly summarized followed by drug-loading techniques to demonstrate how drug release characteristics of polymeric nanoparticles can be manipulated.

8.1 Polymeric Nanoparticle Preparation Methods

Polymeric nanoparticles can be formulated from either preformed polymers or monomers by a variety of methods. However, the most popular methods to prepare polymeric nanoparticles start from biocompatible preformed polymers (Hans and Lowman 2002).

8.1.1 Emulsification Solvent Evaporation Method

The polymer is dissolved in an organic solvent first, and then the drug is dissolved into the polymer solution to create a mixture of drugs dispersed in a polymer solution. The emulsion is prepared by adding water and a surfactant to the polymer solution. In many cases, the nanosized polymer droplets are induced by sonication or homogenization. The organic solvent is then evaporated and the nanoparticles are usually collected by centrifugation and lyophilization (Torche et al. 2000; Suh et al. 1998; Song et al. 1997; Cheng et al. 1998; Feng and Huang 2001). In this method, both water-soluble and water nonsoluble drugs can be incorporated into polymeric nanoparticles; however, hydrophilic drugs have traditionally showed a lower incorporating efficiency (Hans and Lowman 2002).

8.1.2 Emulsification-Diffusion Method

Here, the solvent is chosen as a partially water-soluble solvent (e.g., acetone). The polymer and drug are dissolved in the solvent and emulsified in the aqueous phase containing the stabilizer. The role of the stabilizer is to prevent emulsion droplets from aggregating. Then, water is added to the emulsion to allow for the diffusion of the solvent into the water. The solution is stirred, resulting in the precipitation of nanoparticles which can be collected by centrifugation (Kwon et al. 2001; Takeuchi et al. 2000). The problem with this method is that water-soluble drugs tend to leak out during the diffusion of the solvent.

8.1.3 Nanoprecipitation Method

The polymer and drugs are dissolved in acetone and added to a solution containing Pluronic F68, which is a difunctional block copolymer surfactant terminating in primary hydroxyl groups, a nonionic surfactant that is 100% active and relatively nontoxic. The acetone is then evaporated under reduced pressure and the nanoparticles remain in the suspension (Vergera et al. 1998).

8.1.4 Salting-out Process

Water-in-oil emulsion is formed containing the polymer, acetone, magnesium acetate tetrahydrate, stabilizer, and the drugs. Then, water is added until the volume is sufficient to allow for diffusion of the acetone into the water, which results in the formation of nanoparticles. The suspension is purified by cross-flow filtration and lyophilization (Leroux et al. 1996). One disadvantage to this procedure is its use of the salts that are incompatible with many bioactive compounds.

As mentioned, the above methods are the most popular techniques to prepare nanoparticles from a platform of biocompatible preformed polymers (such as poly (lactic-co-glycolic acid) (PLGA), poly(lactic acid), poly(hexadecylcyanoacrylate), etc.). However, biodegradable nanoparticles can also be made from monomer polymerization methods (Sakuma et al. 2001; Behan et al. 2001). Hydrophilic polysaccharides (like chitosan (CS)) can also be used to make polymeric nanoparticles by the spontaneous ionic gelatin process (Vila et al. 2002; Janes et al. 2001). This technique is becoming attractive because it does not use harmful organic solvents and can formulate particles of small sizes and positive surface potentials.

8.2 Control the Properties of Polymeric Nanoparticles

Along with various methods for forming polymeric nanoparticles is the ability to control the properties of the nanoparticles themselves. First, the polymer and its concentration used in forming nanoparticles strongly affect the particle properties. Besides different hydrophilicity and biocompatibility properties, polymers usually degrade at different rates leading to different drug release characteristics. The molecular weights of polymers have an inverted relation with the particle size, i.e., the smaller nanoparticles can be formulated from lower molecular weight polymers, but also have lower drug encapsulation efficiency in general (Hans and Lowman 2002). However, higher polymer concentrations and higher polymer molecular weights lead to higher encapsulation efficiency and larger sizes of the nanoparticles (Blanco and Alonso 1997; Song et al. 1997; Kwon et al. 2001). For example, Table 2 demonstrates the relationship between molecular weights and concentrations of PLGA on bovine serum albumin (BSA) capture efficiency where PLGA nanoparticles were prepared by emulsion solvent evaporation.

As seen in Table 2, the nanoparticles made from lower molecular weight PLGA or lower PLGA concentrations showed much lower BSA entrapment than those from higher molecular weight PLGA. Figure 10 demonstrates the effect of PLGA concentrations on the size of PLGA nanoparticles prepared by the emulsification-diffusion method using poly(vinyl alcohol) (PVA) as a stabilizer (Kwon et al. 2001). Larger size PLGA nanoparticles were formed with higher PLGA concentrations.

Second, the type and amount of surfactant/stabilizer are also factors that affect properties of nanoparticles as well as drug encapsulation characteristics. For example, phospholipids, a natural emulsifier, have been shown to improve flow and phagocytal properties of dipalmitoyl-phosphatidylcholine (DPPC) due to a more dense packing of DPPC molecules on the surface of the nanoparticles resulting in a smoother surface than particles made with the synthetic polymer, PVA (126). DPPC also improved the encapsulation efficiency compared to PVA using the emulsification solvent evaporation method. Another example is for the case of PLGA nanoparticles. Studies have shown that the PLGA nanoparticles were smaller when formulated using didodecyl dimethyl ammonium bromide (DMAB) than those prepared with PVA (Kwon et al. 2001). In another study, PLGA nanoparticles were formulated using PVA, chitosan, or PVA–chitosan blends as different stabilizers (Ravi et al. 2004). It was shown (Fig. 11) in this study that the sizes and morphologies of PLGA nanoparticles were different for different stabilizers.

Table 2 Effect of PLGA molecular weight (MW) and concentration on bovine serum albumin (BSA) incorporation (adapted with permission from (Song et al. 1997))

MW of PLGA	Concentration of PLGA (%)	Concentration of BSA (%)	Efficiency of BSA capture (%)
58,000	3.0	10.0	24.8
58,000	6.0	14.0	36.8
10,2000	3.0	10.0	68.0
10,2000	6.0	14.0	74.8

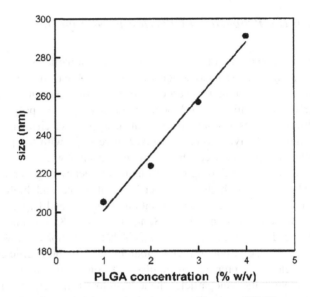

Fig. 10 Effect of PLGA concentration on the mean particle size of PLGA nanoparticles (at the concentration of 2.5 (weigh/volume %) of poly(vinyl alcohol) (PVA)) (Reprinted with permission from (Kwon et al. 2001))

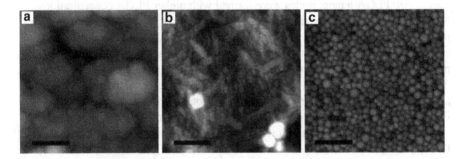

Fig. 11 Atomic force microscopy images of PLGA nanoparticles: (**a**) PLGA nanoparticles with PVA alone as a stabilizer, (**b**) with chitosan alone, and (**c**) with a PVA-chitosan blend. *Bars* represent 150 nm (Adapted with permission from (Ravi et al. 2004)

Third, the zeta potential of polymeric nanoparticles can be controlled. Zeta potential is a measure of the charge of the particle. The larger the absolute value of the zeta potential of the particle, the larger the amount of charge present on its surface. Zeta potential has a strong influence on the stability of nanoparticles as repulsive forces induced between high zeta potential particles lead to the formation of more stable particles with a more uniform size distribution. This stability is important in preventing aggregation. When a surface modification is added (like PEG), the negative zeta potential is lowered, increasing the nanoparticle stability (Vila et al. 2002).

8.3 Drug-Loading Methods

Drugs can be entrapped in a polymer matrix, encapsulated in a nanoparticle core, surrounded by a shell-like polymer membrane, chemically conjugated to the polymer, or bound to the particle's surface by adsorption. The method of drug loading can be divided into two classes: (1) incorporation of the drugs during the formulation of polymeric nanoparticles and (2) absorption of the drugs into nanoparticles after the formulation of the particles. It was shown that the absorption method to load drugs after forming polymeric nanoparticles has lower loading efficiency than incorporating the drugs during the process of forming nanoparticles (Alenso et al. 1991; Ueda et al. 1998).

In the incorporation method, several factors can affect drug-loading efficiency. The concentration of monomer and the concentration of drugs can both determine the drug entrapment efficiency. For example, the drug loading of DOX into poly(butylcyanoacrylate) (PBCA) nanoparticles was shown to strongly depend on the concentration of DOX (Yang et al. 2000). Increasing the concentration of DOX from 5 to 20% resulted in a decrease in the entrapment efficiency from 24.9% to 10.5%.

In the absorption method, the surface properties of nanoparticles (such as hydrophilicity) have a strong impact on drug loading. These surface properties, in turn, are functions of the properties of the polymers themselves and the methods used during preparation.

8.4 Drug Release Characteristics and Drug Biodistribution Profiles

Since polymeric nanoparticles can be fabricated from various polymers by many different methods, the release characteristics of the nanoparticles loaded with drugs can be fine-tuned. By varying the polymer composition of the particle and its morphology, it is possible to fine-tune a variety of controlled-release characteristics, allowing moderate continuous doses over prolonged periods of time (Anderson and Shive 1997). The degradation rate of the polymers and the corresponding drug release rate can vary from days to months and can be easily manipulated by varying the composition of polymers or copolymers (Brigger et al. 2002). For example, polymeric nanoparticles were formulated from PLGA-monomethoxypoly (ethyleneglycol) (mPEG) copolymers of different compositions (PLGA-mPEG molar ratios) and loaded with cisplatin (that is, a platinum-based chemotherapy drug used to treat various types of cancers, including sarcomas, some carcinomas, and lymphomas) (Avgoustakis et al. 2002). The composition of the PLGA-mPEG nanoparticles was shown to affect the drug release profile, i.e., the rate of release increased when the mPEG content of the nanoparticles increased. The zeta potentials and biodegradability of the PLGA-mPEG particles were also influenced by the content of mPEG in the copolymer.

Not only does the polymeric composition of the nanoparticle determine the release profile of the drug, but properties of the drug itself (such as molecular weight, charge, localization in the nanospheres, drug loading (adsorbed or incorporated)) can also greatly influence the drug distribution pattern in the reticuloendothelial organ (Couvreur et al. 1980). For example, it was also demonstrated that the incorporation of cytostatic drugs (drugs that stop cancer cells from multiplying in contrast to cytotoxic drugs that kill cancer cells) into polymer nanoparticle carriers results in a change in the drug biodistribution profile. In one experiment in mice treated with DOX incorporated into poly(isohexylcyanoacrylate) (PIHCA) nanoparticles, higher concentrations of DOX were found in the liver, spleen, and lungs, as compared to the counterpart mice treated with free DOX (Verdun et al. 1990). In another experiment, it was demonstrated that actinomycin D adsorbed on poly(methylcyanoacrylate) (PMCA) nanoparticles concentrated mainly on the lungs of the rats (Brasseur et al. 1980), while actinomycin D incorporated into the more slowly biodegradable poly(ethylcyanoacrylate) (PECA) nanoparticles, the drug accumulated mainly in the small intestine of rats (Couvreur et al. 1980). Even with the same PECA nanoparticles, when loaded with vinblastine (that is, an anticancer drug used to treat certain kinds of cancer, including Hodgkin's lymphoma, nonsmall cell lung cancer, breast cancer, and testicular cancer), the drug was highly concentrated in the spleen of rats (Couvreur et al. 1980).

8.5   Disadvantages of Polymeric Nanoparticles

Although polymeric nanoparticles have many advantages (such as various fabrication methods, versatile drug loading, and drug release methods), they still have some drawbacks. The main limitation is that some preparation methods use toxic organic solvents that could degrade certain drugs when they come into contact during the formulation process or could be hazardous to the environment as well as to the physiological system (Birnbaum et al. 2000). Other disadvantages include poor drug encapsulation for certain hydrophilic drugs and the possible drug leakage before reaching its target tissues and/or cells.

9   Cancer Biosensors

Many of the above materials used to fight cancer have been incorporated into novel in situ sensors that can determine cancer cell presence. A biosensor able to detect cells would be an all-in-one dream device for such applications (de la Escosura-Muniz et al. 2009). Specifically, de la Escosura-Muniz et al. developed an electrocatalytic device for the specific identification of tumor cells that quantifies gold nanoparticles (AuNPs) coupled with an electrotransducing platform/sensor (Fig. 12). Proliferation and adherence of tumor cells were achieved on the electrotransducer/detector, which consists of a mass-produced screen-printed carbon

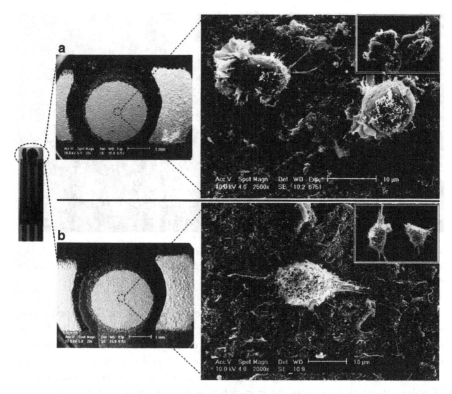

Fig. 12 SEM images of the electrotransducer (SPCE) (*left*) with its three surfaces and details of the (**a**) HMy2 and (**b**) PC-3 cell lines on the carbon working electrode (*right*). Inset images correspond to cell growth on the plastic area of the SPCEs

electrode (SPCE). In situ identification/quantification of tumor cells was achieved with a detection limit of 4,000 cells per 700 μL of suspension. This novel and selective cell-sensing device was based on the reaction of cell surface proteins with specific antibodies conjugated with AuNPs. Final detection required only a couple of minutes, taking advantage of the catalytic properties of AuNPs on hydrogen evolution. The proposed detection method did not require the chemical agents used in most existing assays for the detection of AuNPs. It allowed for the miniaturization of the system and has been reported to be much cheaper than other expensive and sophisticated methods used for tumor cell detection. De la Esocura-Muniz et al. envisaged that this device could operate in as simple of a way as an immunosensor or DNA sensor. Moreover, it could be used, even by inexperienced staff, for the detection of protein molecules or DNA strands.

In recent years, there have been some attempts at cancer cell analysis using optical-based biosensors. In addition to optical biosensors, sensitive electrochemical DNA sensors, immunosensors, and other bioassays have all recently been developed using nanoparticles (NPs) as labels and providing direct detection without prior chemical dissolution.

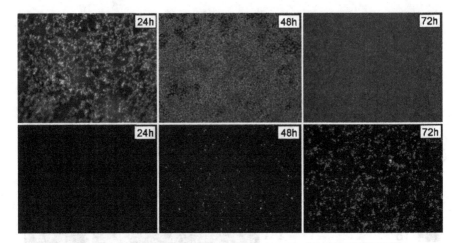

Fig. 13 Fluorescence microscopy images of Hoechst 33342 and propidium iodide-stained MCF-7 cells cultured on the magnetoelastic sensor recorded under ultraviolet (*top*) and *green* light irradiation (*bottom*), respectively

Moreover, wireless cancer cell detection devices have been developed (Xiao et al. 2008). Specifically, a wireless sensing device was developed for the in situ monitoring of the growth of human breast cancer cells (MCF-7) and evaluation of the cytotoxicity of the anticancer drugs fluorouracil and cisplatin (Fig. 13). The sensor was fabricated by coating a magnetoelastic ribbon-like sensor with a layer of polyurethane that protects the iron-rich sensor from oxidation and provides a cell-compatible surface. In response to a time-varying magnetic field, the magnetoelastic sensor longitudinally vibrates, emitting a magnetic flux that can be remotely detected by a pick-up coil. No physical connections between the sensor and the detection system were required. This wireless capability facilitated aseptic biological operation, especially in cell culture as illustrated in this work. The adhesion of cells on the sensor surface resulted in a decrease in the resonance amplitude, which is proportional to the cell concentration. A linear response was obtained in cell concentrations of 5×10^4 to 1×10^6 cells ml^{-1}, with a detection limit of 1.2×10^4 cells ml^{-1}. The adhesion strength of cells on the sensor was qualitatively evaluated by increasing the amplitude of the magnetic excitation field and the cytotoxicity of the anticancer drugs fluorouracil and cisplatin was evaluated by the magnetoelastic biosensor. The cytostatic curve was related with the quantity of cytostatic drug. The lethal concentration (LC50) for cells incubated in the presence of drugs for 20 h was calculated to be 19.9 µM for fluorouracil and 13.1 µM for cisplatin.

10   Conclusions

The research for more sensitive diagnostic methods, more targeted delivery methods, and superior therapeutic agents has been tremendously expanding over the last 10 years. Some advances have been utilized in the clinics, but most of them are still under aggressive developmental stages. The ultimate goal is to lengthen the life of cancer patients by developing the tools that can detect cancer in early stages and then treat them while minimizing side effects. There is undoubtedly a lot more that needs to be studied and improved in order to utilize even just some of the above-mentioned nanoparticles in a clinical setting. To understand the advantages and disadvantages of each type of nanoparticle, intensive research is underway to make them more targeted and less harmful to healthy tissues while still delivering efficient therapeutic agents and imaging capabilities. However, it is clear that with nanotechnology, we are on the verge of developing exciting in situ cancer sensors.

References

Aden, AL; Kerker, M. Scattering of electromagnetic waves from two concentric spheres. *J Appl Phys*, 1951 22, 1242–1246.

Åkerman, ME; Chan, WCW; Laakkonen, P; Bhatia, SN; Ruoslahti, E. Nanocrystal targeting in vivo. *Proceedings of the National Academy of Sciences*, 2002 99, 12617–12621.

Akyildiz, IF; Brunetti, F; Blázquez C. Nanonetworks: A new communication paradigm. *Computer Networks* 2008 52, 2260–2279.

Alenso, MJ; Losa, C; Calvo, P; Vila-Jato, JL. Approaches to improve the association of amikacin sulphate to poly(cyanoacrylate) nanoparticles. *International Journal of Pharmaceutics*, 1991 68, 69–76.

Alexiou, C; et al. Locoregional cancer treatment with magnetic drug targeting. *Cancer Research*, 2000 60, 6641–6648.

Alivisatos, AP. Semiconductor clusters, nanocrystals, and quantum dots. *Science*, 1996 271, 933–937.

American Cancer Society. *Cancer facts and figures – 2008*, ACS, Atlanta, GA, 2008.

Anderson, JM; Shive, MS. Biodegradation and biocompatibility of PLA and PLGA microspheres. *Advanced Drug Delivery Reviews*, 1997 28, 5–24.

Anyarambhatla, GR; Needham, D. Enhancement of the phase transition permeability of DPPC liposomes by incorporation of MPPC: a new temperature-sensitive liposome for use with mild hyperthermia. *Journal of Liposome Research*, 1999 9, 491–506.

Avgoustakis, K, Beletsia, A, Panagia, Z, Klepetsanisa, P, Karydasb, AG, Ithakissios, DS. PLGA-mPEG nanoparticles of cisplatin: in vitro nanoparticle degradation, in vitro drug release and in vivo drug residence in blood properties. *Journal of Controlled Release*, 2002 79, 123–135.

Bailey, RE; Nie, SM. Alloyed semiconductor QDs: tuning the optical properties without changing the particle size. *Journal of the American Chemical Society*, 2003 125, 7100–7106.

Baish, JW; Jain, RK. Fractals and cancer. *Cancer Research*, 2000 60, 3683–3688.

Ballou, B; Lagerholm, BC; Ernst, LA; Bruchez, MP; Waggoner, AS. Noninvasive imaging of quantum dots in mice. *Bioconjugate Chemistry*, 2004 15, 79–86.

Baraton, MI; Chen, X; Gonsalves, KE. FTIR study of nanostructured alumina nitride powder surface: determination of the acidic/basic sites by CO, CO_2, and acetic acid adsorptions. *Nanostructured Matter*, 1997 8, 435.

Behan, N; Birkinshaw, C; Clarke, N. Poly n-butyl cyanoacrylate nanoparticles: a mechanistic study of polymerisation and particle formation. *Biomaterials*, 2001 22, 1335–1344.

Benkoski, JJ; Jesorka, A; Edvardsson, M; Hook, F. Light-regulated release of liposomes from phospholipid membranes via photoresponsive polymer-DNA conjugates. *Soft Matter* 2006 2: 710–715.

Birnbaum, DT; Kosmala, JD; Henthorn, DB; Peppas, LB. Controlled release of β-estradiol from PLAGA microparticles: The effect of organic phase solvent on encapsulation and release. *Journal of Controlled Release*, 2000 65, 375–387.

Blanco, MD; Alonso, MJ. Development and characterization of protein-loaded poly(lactide-co-glycolide) nanospheres. *European Journal of Pharmaceutics and Biopharmaceutics,* 1997 43, 287–294.

Bonadonna, G; et al. Drugs ten years later: epirubicin. *Annals of Oncology,* 1993 4, 359–369.

Brannon-Peppas, L; Blanchette, JO. Nanoparticle and targeted systems for cancer therapy. *Advanced Drug Delivery Reviews* 2004 56, 1649–1659.

Brasseur, F; et al. Actinomycin D adsorbed on polymethylcyanoacrylate nanoparticles: increased efficiency against an experimental tumor. *European Journal of Cancer,* 1980 10, 1441–1445.

Brigger, I; Dubernet, C; Couvreur, P. Nanoparticles in cancer therapy and diagnosis. *Advanced Drug Delivery Reviews,* 2002 54, 631–651.

Brown, MA; Semelka, RC. *MRI: Basic principles and applications.* 2nd ed. New York, NY: Wiley-Liss, 1999.

Carmeliet, P; Jain, RK. Angiogenesis in cancer and other diseases. *Nature* 2000 407, pp 249–257.

Chang, SS; Reuter, VE; Heston, WDW; Gaudin, PB. Metastatic renal cell carcinoma neovasculature expresses prostate-specific membrane antigen. *Urology,* 2001 57, 801–805.

Chatterjee, J; Bettge, M; Haik, Y; Chen, CJ. Synthesis and characterization of polymer encapsulated Cu–Ni magnetic nanoparticles for hyperthermia applications. *Journal of Magnetism and Magnetic Materials,* 2005 293, 303–309.

Chen, AM; Scott, MD. Current and future applications of immunologic attenuation via PEGylation of cells and tissues. *BioDrugs,* 2001 15, 833–847.

Chen, C; Hayek, S; Mohite, V; Yuan, H; Chatterjee, J; Haik, Y. Title: Nanomagnetics and magnetic hyperthermia. In: Nalwa HS, Webster TJ editors. *Cancer nanotechnology: Nanomaterials for cancer diagnosis and therapy.* San Diego, CA: American Scientific, 2007, pp 160–191.

Cheng, YH; Illum, L; Davis, SS. A poly(image-lactide-co-glycolide) microsphere depot system for delivery of haloperidol. *Journal of Controlled Release,* 1998 55, 203–212.

Couvreur, P; Kante, B; Lenaerts, V; Scailteur, V; Roland, M; Speiser, P. Tissue distribution of antitumor drugs associated with polyalkylcyanoacrylate nanoparticles. *Journal of Pharmaceutical Sciences,* 1980 69, 199–202.

de la Escosura-Muniz, A; Sanchez-Espinel, C; Diaz-Freitas, B; Gonzalez-Fernandez, A; Maltez-da Costa, M; Merko, A. Rapid identification and quantification of tumor cells using an electrocatalytic method based on gold nanoparticles. *Anal Chem,* 2009, 81(24), 10268–10274.

Derfus, AM; et al. Probing the cytotoxicity of semiconductor quantum dots. *Nano Letters,* 2004 4, 11–18.

Drummond, DC; Meyer, O; Hong, K; Kirpotin, DB; Papahadjopoulos, D. Optimizing liposomes for the delivery of chemotherapeutic agents to solid tumors. *Pharmacological Reviews,* 1999 51, 691–744.

Dubertret, B; Skourides, P; Norris, DJ; Noireaux, V; Brivanlou, AH; Libchaber, A. In vivo imaging of quantum dots encapsulated in phospholipid micelles. *Science,* 2002 298, 1759–1762.

Eatock, MM; Schätzlein, A; Kaye, SB. Tumour vasculature as a target for anticancer therapy. *Cancer Treatment Review,* 2000 26, 191–204.

Efros, AL; Rosen, M. The electronic structure of semiconductor nanocrystals. *Annual Review of Materials Science,* 2000 30, 475–521.

Fairbairn, JJ; Khan, MW; Ward, KJ; Loveridge, BW; Fairbairn, DW; O'Neill, KL. Induction of apoptotic cell DNA fragmentation in human cells after treatment with hyperthermia. *Cancer Letters,* 1995 89, 183–188.

Falk, MH; Issels RD. Hyperthermia in oncology. *International Journal of Hyperthermia* 2001 17, 1–18.

Feng, SS; Huang, G. Effects of emulsifiers on the controlled release of paclitaxel (Taxol) from nanospheres of biodegradable polymers. *Journal of Controlled Release,* 2001 71, 53–69.

Ferrari, M. Cancer nanotechnology: opportunities and challenges. *Nature Reviews,* 2005 5, 161–171.

Gabizon, A; Shmeeda, H; Horowitz, AT; Zalipsky, S. Tumor cell targeting of liposome-entrapped drugs with phospholipid-anchored folic acid-PEG conjugates. *Advanced Drug Delivery Reviews*, 2004 56, 1177–1192.

Gao, XH; Nie, SM. Doping mesoporous materials with multicolor quantum dots. *Journal of Physical Chemistry B*, 2003 107, 11575–11578.

Gao, X; Cui, Y; Levenson, RM; Chung, LW; Nie, S. In vivo cancer targeting and imaging with semiconductor quantum dots. *Nature Biotechnology*, 2004a 22, 969–976.

Gao, XH; Nie, SM. Quantum dot-encoded mesoporous beads with high brightness and uniformity: rapid readout using flow cytometry. *Annals of Chemistry*, 2004b 76, 2406–2410.

Gao, X; Yang, L; Petros, JA; Marshall, FF; Simons, JW; Nie, S. In vivo molecular and cellular imaging with quantum dots. *Current Opinion in Biotechnology*, 2005 16, 63–72.

Garin-Chesa, P; Campbell, I; Saigo, P; Lewis, J; Old, L; Rettig, W. Trophoblast and ovarian cancer antigen LK26. Sensitivity and specificity in immunopathology and molecular identification as a folate-binding protein. *American Journal of Pathology*, 1993 142, 557–567.

Gerner, EW; et al. Heat-inducible vectors for use in gene therapy. *International Journal of Hyperthermia*, 2000 16, 171–181.

Giaever, I. Nanotechnology, biology, and business. *Nanomedicine: Nanotechnology, Biology and Medicine*, 2006 2, 268.

Grossfeld, GD; Carrol, PR; Lindeman, N. Thrombospondin-1 expression in patients with pathologic state T3 prostate cancer undergoing radical prostatectomy: Association with p53 alterations, tumor angiogenesis and tumor progression, *Urology*, 2002 59, 97–102.

Grundemann, M; et al. Ultranarrow luminescence lines from single quantum dots. *Physical Review Letters*, 1995 74, 4043–4046.

Hafez, IM; Ansell, S; Cullis, PR. Tunable pH-sensitive liposomes composed of mixtures of cationic and anionic lipids. *Biophysical Journal*, 2000 79, 1438–1446.

Han, MY; Gao, XH; Su, JZ; Nie, SM. Quantum dot-tagged microbeads for multiplexed optical coding of biomolecules. *Nature Biotechnology*, 2001 19, 631–635.

Hans, ML; Lowman, AM. Biodegradable nanoparticles for drug delivery and targeting. *Current Opinion in Solid State and Materials Science*, 2002 6, 319–327.

Hardman, R. A Toxicologic review of quantum dots: Toxicity depends on physicochemical and environmental factors. *Environmental Health Perspectives*, 2006 114, 165–172.

Heath, JR; Kuekes, PJ; Snider, GS; Williams, RS. A defect-tolerant computer architecture: Opportunities for nanotechnology. *Science* 1998 280, 1716–1721.

Helmlinger, G; Yuan, F; Dellian, M; Jain, RK. Interstitial pH and pO_2 gradients in solid tumors in vivo: high-resolution measurements reveal a lack of correlation. *Nature Medicine*, 1997 3, 177–182.

Hirsch, LR; et al. Nanoshell mediated near-infrared thermal therapy of tumors under magnetic resonance guidance. *Proceedings of the National Academy of Sciences*, 2003 100, 13549–13554.

Hobbs, SK; Monsky, WL; Yuan, F; Roberts, WG; Griffith, L; Torchilin, VP; Jain, RK. Regulation of transport pathways in tumor vessels: Role of tumor type and microenvironment. *Proceedings of the National Academy of Sciences*, 1998 95, 4607–4612.

Hood, JD; et al. Tumor regression by targeted gene delivery to the neovasculature. *Science*, 2002 296, 2404–2407.

Huang, S; MacDonald, RC. Acoustically active liposomes for drug encapsulation and ultrasound-triggered release. *Biochimica et Biophysica Acta (BBA) – Biomembranes*, 2004 1665, 134–141.

Jain, KK. Nanodiagnostics: Application of nanotechnology in molecular diagnostics. *Expert Review of Molecular Diagnostics*, 2003 3, 153–161.

Janes, KA; Calvo, P; Alonso, MJ. Polysaccharide colloidal particles as delivery systems for macromolecules. *Advanced Drug Delivery Reviews*, 2001 47, 83–97.

Jesorka, A; Orwar, O. Liposomes: Technologies and analytical applications. *Annual Review of Analytical Chemistry*, 2008 1, 801–832.

Jones, A; Harris, AL. New developments in angiogenesis: a major mechanism for tumor growth and target for therapy. *Cancer Journal from Scientific American*, 1998 4, 209–217

Jordan, A; et al. effects of magnetic fluid Hyperthermia (MFH) on C3H mammary carcinoma in vivo. *International Journal of Hyperthermia*, 1996 12, 587–605.

Jordan, A; Scholz, R; Wust, P; Fahling, H; Felix, R. Magnetic fluid hyperthermia (MFH): Cancer treatment with AC magnetic field induced excitation of biocompatible superparamagnetic nanoparticles. *Journal of Magnetism and Magnetic Materials*, 1999 201, 413–419.

Kapp, DS; Hahn, GM; Carlson, RW. Title: Principles of hyperthermia. In: Bast RCJ, et al. editors. *Cancer Medicine* 5th ed. Hamilton, ON: B.C. Decker, 2000.

Kirpotin, D; Hong, K; Mullah, N; Papahadjopoulos, D; Zalipsky, S. Liposomes with detachable polymer coating: destabilization and fusion of dioleoylphosphatidylethanolamine vesicles triggered by cleavage of surface-grafted poly(ethylene glycol). *Federation of Biochemical Societies Letters*, 1996 388, 115–118.

Klabunde, KJ; et al. Nanocrystals as stoichiometric reagents with unique surface chemistry. *Journal of Physical Chemistry*, 1996 100, 12141–12153.

Klibanov, AL; Maruyama, K; Torchilin, VP Huang, L. Amphipathic polyethyleneglycols effectively prolong the circulation time of liposomes. *Federation of Biochemical Societies Letters*, 1990 268, 235–237.

Kong, DF; Goldschmidt-Clermont, PJ. Tiny solutions for giant cardiac problems. *Trends in Cardiovascular Medicine,* 2005 15, 207–211.

Kwon, HY; et al. Preparation of PLGA nanoparticles containing estrogen by emulsification-diffusion method. *Colloids and Surfaces A,* 2001 182, 123–130.

Larson, DR; et al. Water-soluble quantum dots for multiphoton fluorescence imaging in vivo. *Science,* 2003 30, 1434–1436.

Leach, MO. Title: Spatially localised nuclear magnetic resonance. In: Webb, S editor. *The physics of medical imaging*. Bristol: Adam Hilger, 1988, pp 389–487.

Leroux, J; et al. Biodegradable nanoparticles-from sustained release formulations to improved site specific drug delivery. *Journal of Controlled Release* 1996 39, 339–350.

Liu, D; Huang, L. Interaction of pH-sensitive liposomes with blood components. *Journal of Liposome Research,* 1994 4, 121–141.

Loo, C; et al. Nanoshell-enabled photonics-based imaging and therapy of cancer. *Technology in Cancer Research & Treatment,* 2004 3, 33–40.

Loo, C;, Lowery, A; Halas, N; West, J; Drezek, R. Immunotargeted nanoshells for integrated cancer imaging and therapy. *Nano letters,* 2005 5, 709–711.

Low, PS; Antony, AC. Folate receptor-targeted drugs for cancer and inflammatory diseases. *Advanced Drug Delivery Reviews,* 2004 56, 1055–1058.

Lubbe, AS; et al. Preclinical experiences with magnetic drug targeting: Tolerance and efficacy. *Cancer Research,* 1996a 56, 4694–4701.

Lubbe, AS; et al. Clinical experiences with magnetic drug targeting: A phase I study with 4'-Epidoxorubicin in 14 patients with advanced solid tumors. *Cancer Research* 1996b 56, 4686–4693.

Lubbe, AS; Bergemann, C; Brock, J; McClure, DG. Physiological aspects in magnetic drug-targeting. *Journal of Magnetism and Magnetic Materials,* 1999 194, 149–155.

Michalet, X; et al. Quantum dots for live cells, in vivo imaging, and diagnostics. *Science,* 2005 307, 538 – 544.

Miller, D; Webster, TJ. Title: Anticancer orthopedic implants. In: Nalwa H, Webster T J, editors. *Cancer nanotechnology* California: American Scientific Publishers, 2007, pp 307–316.

Mizoue, T; et al. Targetability and intracellular delivery of anti-BCG antibody-modified, pH-sensitive fusogenic immunoliposomes to tumor cells. *International Journal of Pharmaceutics,* 2002 237, 129–137.

Moghimi, SM; Hunter, AC. Recognition by macrophages and liver cells of opsonized phospholipid vesicles and phospholipid headgroups. *Pharmaceutical Research,* 2001 18, 1–7.

Mornet, S; Vasseur, S; Grasset, F; Duguet, E. Magnetic nanoparticle design for medical diagnosis and therapy. *Journal of Materials Chemistry,* 2004 14, 2161–2175.

Moroz, P; Jones, SK; Gray, BN. Status of hyperthermia in the treatment of advanced liver cancer. *Journal of Surgical Oncology,* 2001 77, 259–269.

Moroz, P; Jones, SK; Gray, BN. Magnetically mediated hyperthermia: current status and future directions. *International Journal of Hyperthermia,* 2002 18, 267–84.

Nie, S; Xing, Y; Kim, GJ; Simons, JW. Nanotechnology applications in cancer. *Annual Reviews– Biomedical Engineering,* 2007 9, 12.1–12.32.

Niemeyer, CM. Nanoparticles, proteins, and nucleic acids: biotechnology meets materials science. *Angewandte Chemie International Edition,* 2001 40, 4128–4158.

Oldenburg, SJ; Jackson, JB; Westcott, SL; Halas, NJ. Light scattering from dipole and quadrupole nanoshell antennas. *Applied Physics Letters,* 1999 75, 2897–2899.

Orive, G; Hernández, RM; Gascón, RA; Pedraz, JL. Micro and nano drug delivery systems in cancer therapy. *Cancer Therapy,* 2005 3, 131–138.

Overgaard, K; Overgaard, J. Investigations on the possibility of a thermic tumor therapy. *European Journal of Cancer,* 1972 8, 65–78.

Pan, XQ; Wang, H; Lee, RJ. Antitumor activity of folate receptor-targeted liposomal Doxorubicin in a KB oral carcinoma murine xenograft model. *Pharmaceutical Research,* 2003 20, 417–422.

Park, K. Nanotechnology: What it can do for drug delivery. *Journal of Controlled Release,* 2007 120, 1–3.

Park, JW; et al. Anti-HER2 immunoliposomes: enhanced efficacy attributable to targeted delivery. *Clinical Cancer Research,* 2002 8, 1172–1181.

Pastorino, F; et al. Doxorubicin-loaded Fab' fragments of antidisialoganglioside immunoliposomes selectively inhibit the growth and dissemination of human neuroblastoma in nude mice. *Cancer Research,* 2003 63, 86–92.

Perla, V; Webster, TJ. Better osteoblast adhesion on nanoparticulate selenium- A promising orthopedic implant material. *Journal of Biomedical Materials Research A,* 2005, 75, 356–364.

Prodan, E; Radloff, C; Halas, NJ; Nordlander, P. A hybridization model for the plasmon response of complex nanostructure. *Science,* 2003 17, 419–422.

Ran, Y; Yalkowsky, SH. Halothane, a novel solvent for the preparation of liposomes containing 2-4'-amino-3'-methylphenyl benzothiazole (AMPB), an anticancer drug: A technical note. *AAPS PharmSciTech,* 2003 4, E20.

Ran, S; et al. Increased exposure of anionic phospholipids on the surface of tumor blood vessels. *Cancer Research,* 2002 62, 6132–6140.

Ravi, MNV; Bakowsky, U; Lehr, CM. Preparation and characterization of cationic PLGA nanospheres as DNA carriers. *Biomaterials,* 2004 25, 1771–1777.

Reddy, JA; Low, PS. Enhanced folate receptor mediated gene therapy using a novel pH-sensitive lipid formulation. *Journal of Controlled Release,* 2000 64, 27–37.

Reiss, P; Bleuse, J; Pron, A. highly luminescent CdSe/ZnSe core/shell nanocrystals of low size dispersion. *Nano Letters,* 2002 2, 781–784.

Rikans, LE; Yamano, T. Mechanism of cadmium-mediated acute hepatotoxicity. *Journal of Biochemical and Molecular Toxicology,* 2000 14, 110–117.

Rosenthal, SJ; et al. Targeting cell surface receptors with ligand-conjugated nanocrystals. *Journal of the American Chemical Society,* 2002 124, 4586–4594.

Ruddon, RW. *Cancer Biology.* 2nd edition. New York: Oxford University Press, 1987.

Saini, V; et al. Combination of viral biology and nanotechnology: new applications in nanomedicine. *Nanomedicine: Nanotechnology, Biology and Medicine* 2006 2, 200–206.

Sakuma, S; Hayashi, M; Akashi, M. Design of nanoparticles composed of graft copolymers for oral peptide delivery. *Advanced Drug Delivery Reviews,* 2001 47, 21–37.

Sellins, KS; Cohen, JJ. Hyperthermia induces apoptosis in thymocytes. *Radiant Research,* 1991 126, 88–95.

Shantesh, H; Nagraj, H. Nano: The new nemesis of cancer. *Journal of Cancer Research and Therapeutics,* 2006 2, 186–195.

Shellman, YG; et al. Hyperthermia induces endoplasmic reticulum-mediated apoptosis in melanoma and non-melanoma skin cancer cells. *Journal of Investigative Dermatology,* 2008 128, 949–956.

Siegel, RW; Fougere, GE. Mechanical properties of nanophase metals. *Nanostructured Materials,* 1995 6, 205–216

Silva, GA. Introduction to nanotechnology and its applications to medicine. *Surgical Neurology*, 2004 61, 216–220.

Sledge, G; Miller, K. Exploiting the hallmarks of cancer: the future conquest of breast cancer. *European Journal of Cancer*, 2003 39, 1668–1675.

Song, CX; et al. Formulation and characterization of biodegradable nanoparticles for intravascular local drug delivery. *Journal of Controlled Release*, 1997 43, 197–212.

Soppimath, KS; et al. Biodegradable polymeric nanoparticles as drug delivery devices. *Journal of Controlled Release*, 2001 70, 1–20.

Storm, G; Belliot, SO; Daemen, T; Lasic, D. Surface modification of nanoparticles to oppose uptake by the mononuclear phagocyte system. *Advanced Drug Delivery Reviews*, 1995 17, 31–48.

Stroh, M; et al. Quantum dots spectrally distinguish multiple species within the tumor milieu in vivo. *Nature Medicine*, 2005 11, 678–682.

Suh, H; et al. Regulation of smooth muscle cell proliferation using paclitaxel-loaded poly(ethylene oxide)-poly(lactide/glycolide) nanospheres. *Journal of Biomedical Materials Research*, 1998 42, 331–338.

Sun, S; et al. Monodisperse MFe_2O_4 (M=Fe, Co, Mn) nanoparticles. *Journal of the American Chemical Society*, 2004 126, 273–279.

Takeuchi, H; Yamamoto, H; Kawashima, Y. Mucoadhesive nanoparticulate systems for peptide drug delivery. *Advanced Drug Delivery Reviews*, 2001 47, 39–54.

Tang, L; Liu, L; Elwing, H. Complement activation and inflammation triggered by model biomaterial surfaces. *Journal of Biomedical Materials Research*, 1998 41, 333–340.

Teicher, BA. Molecular targets and cancer therapeutics: discovery, development and clinical validation. *Drug Resistance Updates*, 2000 3, 67–73.

Tiwari, SB; Amiji, MM. A review of nanocarrier-based CNS delivery systems. *Current Drug Delivery*, 2006 3, 219–232.

Torche, AM; et al. Ex vivo and in situ PLGA microspheres uptake by pig ileal Peyer's patch segment. *International Journal of Pharmaceutics*, 2000 201, 15–27

Torchilin, VP. Recent advances with liposomes as pharmaceutical carriers. *Nature Reviews Drug Discovery*, 2005 4, 145–160.

Ueda, M; Iwara, A; Kreuter, J. Influence of the preparation methods on the drying release behavior of loperamide-loaded nanoparticles. *Journal of Microencapsulation*, 1998 15, 361–372.

Verdun, C; Brasseur, F; Vranckx, H; Couvreur, P; Roland, M. Tissue distribution of doxorubicin associated with polyhexylcyanoacrylate nanoparticles. *Cancer Chemotherapy and Pharmacology*, 1990 26, 13–18.

Vergera, ML; Fluckigera, L; Kimb, Y; Hoffmana, M; Maincent, P. Preparation and characterization of nanoparticles containing an antihypertensive agent. *European Journal of Pharmaceutics and Biopharmaceutics*, 1998 46, 137–143.

Vila, A; et al. Design of biodegradable particles for protein delivery. *Journal of Controlled Release*, 2002 78, 15–24.

Wang, YX; Hussain, SM; Krestin, GP. Superparamagnetic iron oxide contrast agents: physicochemical characteristics and applications in MR imaging. *European Journal of Radiology*, 2001 11, 2319–2331.

Webster, TJ; Ejiofor, JU. Increased osteoblast adhesion on nanophase metals: Ti, Ti6Al4V, and CoCrMo. *Biomaterials*, 2004 25, 4731–4739.

Webster, TJ; Siegel, RW; Bizios, R. Osteoblast adhesion on nanophase ceramics. *Biomaterials*, 1999 20, 1221–1224

Webster, TJ; Ergun, C; Doremus, RH; Siegel, RW; Bizios, R. Specific proteins mediate enhanced osteoblast adhesion on nanophase ceramics. *Journal of Biomedical Material Research*, 2000a 51, 475–480.

Webster, TJ; Siegel, RW; Bizios, R. Enhanced functions of osteoblasts on nanophase ceramics. *Biomaterials*, 2000b 21, 803–1809.

Webster, TJ; Schadler, LS; Siegel, RW; Bizios, R. Mechanisms of enhanced osteoblast adhesion on nanophase alumina involve vitronectin. *Tissue Engineering*, 2001 7, 291–301.

Weissleder, R. A clearer vision for in vivo imaging. *Nature Biotechnology,* 2001 19, 316–317.

West, JL; Halas, NJ. Engineering nanomaterials for biophotonics applications: Improving sensing, imaging, and therapeutics. *Annual Review of Biomedical Engineering,* 2003 5, 285–292.

Wu, SJ; Jonghe, LC; Rahaman, MN. Sintering of nanophase γ-Al$_2$O$_3$ powder. *Journal of the American Ceramic Society,* 1996 8, 2207–2211.

Wu, X; et al. Immunofluorescent labeling of cancer marker Her2 and other cellular targets with semiconductor quantum dots. *Nature Biotechnology,* 2002 21, 41 – 46.

Wust, P; et al. Hyperthermia in combined treatment of cancer. *Lancet Oncology,* 2002 3, 487–479.

Xiao X, Guo M, Li Q, Cai Q, Yao S, Grimes CA. In-situ monitoring of breast cancer cell (MCF-7) growth and quantification of the cytotoxicity of anticancer drugs fluorouracil and cisplatin. *Biosens Bioelectron.* 2008, 24(2), 247–252.

Yamada, Y; et al. Mitochondrial delivery of mastoparan with transferrin liposomes equipped with a pH-sensitive fusogenic peptide for selective cancer therapy. *International Journal of Pharmaceutics,* 2005 303, 1–7.

Yang, SC; Ge, HX; Hu, Y; Jiang, XQ; Yang, CZ. Doxorubicin-loaded poly(butylcyanoacrylate) nanoparticles produced by emulsifier-free emulsion polymerization. *Journal of Applied Polymer Science,* 2000 78, 517–526.

Yokoyama, M. Drug targeting with nano-sized carrier systems. *International Journal of Artificial Organs,* 2005 8, 77–84.

Yuan, F; et al. Vascular permeability in a human tumor xenograft: Molecular size dependence and cutoff size. *Cancer Research,* 1995 55, 3752–3756.

Zalipsky, S; et al. New detachable poly(ethylene glycol) conjugates: cysteine-cleavable lipopolymers regenerating natural phospholipid, diacyl phosphatidylethanolamine. *Bioconjugate Chemistry,* 1999 10, 703–707.

Chapter 2
Monitoring Tissue Healing Through Nanosensors

Lei Yang and Thomas J. Webster

Abstract Nanotechnology is the use of materials with at least one dimension less than 100 nm. Nanotechnology has already revolutionized numerous fields, from construction to computers. Recently, nanotechnology has also been used to improve disease detection and treatment by developing wireless in situ sensors. Importantly, the use of wireless technologies in medicine, such as wireless body area networks and wireless personal area networks, is not new as they provide many promising applications in medical monitoring systems to measure physiological data from specific anatomical areas. Nanotechnology can aid in the functioning of wireless medical devices since it can provide for materials smaller in size (thus, minimally interacting with tissues to invoke an immune response), better in properties (such as electronic), and more similar to those of natural tissues since natural tissues are composed of nanoscale entities. In fact, studies have demonstrated increased tissue growth, decreased inflammation, and decreased infection of numerous nanoscale compared to currently used micron-scale materials. Due to the above, an ever-expanding range of therapeutic and diagnostic applications are being pursued by academic and industrial researchers. This chapter aims to provide a comprehensive review of recent developments in wireless sensor nanotechnology for monitoring and controlling cell responses.

Keywords Wireless • Nanotechnology • Sensors • Diagnosis • Treatment • Diseases

1 Introduction

It is widely recognized that the growing global population is aging and this trend will place an increasing demand on medical and healthcare resources. Specifically, in 2008, there were more than 650 million people over the age of 65 and this number will double over the next 10 years. In the United States, about 20% of the population

T.J. Webster (✉)
School of Engineering, Brown University, 182 Hope Street, Providence, RI 02917, USA
e-mail: thomas_webster@brown.edu

T.J. Webster (ed.), *Nanotechnology Enabled In situ Sensors for Monitoring Health*,
DOI 10.1007/978-1-4419-7291-0_2, © Springer Science+Business Media, LLC 2011

will be over 65 by 2030, compared to only 12% today. As a consequence of this aging, a number of chronic age-related diseases (specifically, Type 2 diabetes, cancer, congestive heart failure, chronic obstructive pulmonary disease, arthritis, osteoporosis, and dementia) have significantly increased (Fass 2007). In addition, there are more than one billion adults worldwide today who are overweight. Clearly, there is an ever-increasing shortage of doctors and nurses, and an increasing demand for healthcare services (Feied et al. 2006). Such an unbalance between "supply and demand" in medicine will place an enormous strain on our medical communities unless we develop new, novel, medical care technologies.

Some believe the future of medicine resides in wireless sensor technologies. Such wireless medical technologies may be more efficient and effective than today's medical practices. As an example of the importance of wireless medical technologies, today, over 50% of hospitals in the United States have wireless local area networks (WLANs) and widely accessible Wi-Fi and WiMax devices, enabling practitioners to access patient medical information, both at the point of care and anywhere else it is needed (Hao and Foster 2008). Similarly, tablet PCs, PDAs and laptops connected to the WLAN allow clinicians to immediately record medical information in an electronic format, as well as order tests and prescribe medication at the patient's bedside, all from their chosen device (Hao and Foster 2008). It is now commonplace to find doctors with computers (rather than old-fashion paper folders) when meeting with patients.

Inside the body, the use of wireless medical devices is also undergoing a revolution. Proponents of wireless medical devices highlight the possibility to now develop small, low-power, lightweight, and intelligent physiological monitoring devices. Wireless body sensor networks (WBSN) have been developed to provide real time information concerning patient health (Hao and Foster 2008). This chapter will cover some of the more exciting advances in this electronic age where wireless medical devices are revolutionizing medicine. Nanotechnology, or the use of materials with one dimension less than 100 nm, is playing a large part in this revolution by allowing for the design of materials with unique properties to interface with tissues and cells.

2   Wireless Medical Monitor Advantages and Disadvantages: The Concept

There are two essential categories of wireless medical devices: (1) wireless medical monitors and (2) wireless medical devices. Differences between these categories involve whether the devices only monitor health or also treat a medical problem. Wireless physiological measurements have a number of advantages over wired measurements, including ease of use, reduced risk of infection, reduced risk of failure, reduced user discomfort, enhanced mobility, and lower cost of care delivery

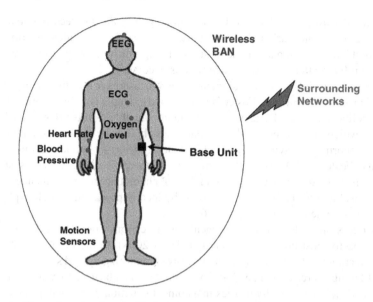

Fig. 1 An illustration of various bodily functions measured by wireless medical monitors

(Townsend et al. 2005) (Fig. 1). Many of these advantages come from the fact that no invasive procedure is needed to obtain a diagnosis using wireless medical monitors.

The paramount advantage of wireless measurements and, thus, wireless medical devices, is their potential to treat a medical problem at the location it was measured (Hao and Foster 2008). Moreover, the use of wireless physiological measurement systems outside the hospital could also help to reduce overall healthcare costs, especially among patients with chronic diseases capable of receiving care at home. Of course, the main disadvantage of wireless medical technology is the increased challenges in designing and fabricating wireless medical devices. Wireless physiological measurements have not only found applications in healthcare, but have also been applied in the areas of military, security, sport, and fitness monitoring (Hao and Foster 2008).

Clearly, a wireless physiological measurement system is meant to alert the medical emergency system if vital signs drop below certain threshold (Hao and Foster 2008). In this scenario, the exact location of the patient needs to be transmitted, along with any useful medical information that could assist the emergency team. For example, in the United States, each year, about 1.1 million Americans suffer a heart attack. About 460,000 of those heart attacks are fatal. About half of those deaths occur within 1 h of the start of symptoms and before the person reaches the hospital (Hao and Foster 2008). The use of wireless physiological measurement systems could help save lives, in that they could detect and warn of early symptoms of impending cardiac (or other) problems, enabling the patient to receive potentially life-saving treatments earlier (Hao and Foster 2008). Of course, identification of a health problem is only part of the solution. Designing sensors that can treat the medical problem after it has been identified is necessary.

A wireless physiological measurement system measures in real-time, a bio-signal for local processing (Hao and Foster 2008). A good example is an automatic internal cardiac defibrillator (AICD, also known as an implantable cardioverter defibrillator or ICD), which acts to restore the regular heart rhythm by delivering an electric shock if abnormal behavior is detected, potentially averting sudden cardiac death (Hao and Foster 2008). Another example is implantable drug delivery systems, which deliver medication more efficiently for chemotherapy, pain management, diabetic insulin delivery, and AIDS therapy, by locally processing wireless physiological measurements (Jones et al. 2006).

A wireless physiological measurement system can also provide real bio-signal information for post-processing (Hao and Foster 2008). In the military, a WPMS implementation can facilitate remote noninvasive monitoring of vital signs of soldiers during training exercises and combat. For example, it can be used to remotely determine a soldiers' condition by medics in a combat situation, without exposing first-responders to increased risks, or to quickly identify the severity of injuries and continuously track the injured condition until they arrive safely at a medical care facility (Hao and Foster 2008). Such devices can keep track of an injured person's vital signs, allowing rapid distribution of the information to medical providers and assisting emergency responders in making critical, and often life-saving, decisions in order to expedite rescue operations (Mendelson et al. 2006). Examples of wireless physiological measurement systems can be found in Table 1 while those being used commercially and researched academically can be found in Tables 2 and 3, respectively (Hao and Foster 2008).

Table 1 Applications of wireless physiological measurement systems and their classification

Critical monitoring	Noncritical monitoring
Monitoring chronically ill patients with heart disease, diabetes, and epilepsy	Monitoring physical conditions and efficiency of a sport athlete during exercises
Monitoring at home and nursing home for elderly and demented people	Control and feedback during athlete training
Monitoring vital signs of soldiers in battle	Crime investigation with wireless lie detectors
Vehicles such as ambulances when transporting patients	In the hospital to reduce discomfort and restriction of wires
Monitoring the consciousness of drivers, pilots, and operators of heavy machinery	Monitoring employees to identify those who are engaged in unlawful activities
Medical research teams can carry out unobtrusive patient studies and clinical field trials over an extensive period	
Remote telemedicine	

Table 2 Some current commercial applications of wireless physiological measurement systems

Commercial applications/vendor	Description	Market
TeleMuse Biocontrol Systems	This is a mobile physiological monitor for acquiring ECG, EMG, EOG, EEG, and GSR data from wireless sensors using ZigBee technology	Medical care and research
VitalSense Integrated Physiological Monitoring System	This is a chest-worn wireless physiological monitor that incorporates an ECG-signal processor and offers wireless transmission of heart rate and respiration rate to a handheld monitor	Fitness and exercise
The Security Alert Tracking System, Third Eye Inc.	Wrist-mounted surveillance monitors blood oxygen saturation and heart rate fluctuations noninvasively; the information is transmitted wirelessly to a central monitoring system. It can assist in apprehending employees engaged in unlawful activities in casino and banks	Security and safety
The Alive Heart and Activity Monitor, Alive	This Bluetooth device monitors the heart rate and activity, including ECGs, blood oximeters, and blood glucose meters. It communicates with software on your mobile phone to log and upload information to a central Internet server	Medical care, research, fitness, and exercise
Polar Heart Rate Monitor/Watch S625X Polar	This is a watch combined with a heart rate monitor, altimeter, and speed/distance monitor. It communicates wirelessly with a chest belt	Fitness and exercise
PillCam® Capsule Endoscopy Given Imaging	The tiny camera contained in the capsule captures images of the gastrointestinal (GI) tract as it travels through the body and transmits the images to a computer, so the physician can view them and make a diagnosis	Medical care

3   Implantable Wireless Medical Devices

3.1   Treating Bone Defects

As mentioned above, while monitoring human health through the use of wireless technologies is important, so is using that information to treat a medical problem (should it be diagnosed). A great example of this approach is through the use of implantable wireless medical devices for diagnosing and treating orthopedic problems. In the United States, annually an estimated 1.5 million people suffer a bone fracture caused by various bone diseases, resulting in 165,000 hip joints and 326,000 knees replacements in 2001 (Smith 2004; Bren 2004). The number of orthopedic implant surgeries is increasing. For example, according to the American

Table 3 Examples of on-going academic research on wireless physiological measurement systems

Research applications/vendor	Description	Market
CodeBlue: wireless sensor networks for medical care, Harvard University	Exploring applications of wireless sensor network technology to raise alerts when the vital signs of patients fall outside the normal range	Medical care/ military
Wireless physiological sensors for ambulatory and implantable applications, Tampere University of Technology	The study and development of a new wireless sensor technology for ambulatory and implantable human psychophysiological applications. The goal is to develop commercially mass-produced physiological measurement systems, based on patch-type sensors and implantable smart wireless devices	Medical care, research, and military
Wireless implantable sensors with advanced on-body data processing, Queen Mary College, University of London	The proposed feasibility study aims to deliver a clinically viable strategy that can provide a wireless connected system for implantable electrophysiological and metabolic monitoring sensors, enhancing existing capabilities in both wireless and sensor technology	Medical care

Academy of Orthopedic Surgeons, there was an 83.72% increase in the number of hip replacements performed from nearly 258,000 procedures in 2000 to 474,000 procedures (including 234,000 total and 240,000 partial hip replacements) in 2004. The total hospitalization costs for knee replacements doubled to $11.38 billion in 2003 compared with $5.67 billion in 1999 (American Academy of Orthopedic Surgeons 2010). Despite the increasing demand and cost of orthopedic implants, the durability of implants has not risen as most of the current implants serve only 10–15 years (Webster 2003).

All of these statistics bring a challenge of developing durable, affordable, and better orthopedic implants to the bone community, but the question is how? And, can the design and use of wireless orthopedic medical devices help? Integrating materials science and bone biology is probably one of the best solutions and most importantly, can create better interfaces between the implant and bone. More excitingly, the emergence of nanomaterials (materials with size scales within 1–100 nm, i.e., 10^{-9}–10^{-8} m) and nanotechnology (the research and development related to nanoscale materials) during the last two decades has brought more anticipation to solve the chronic challenge of creating better bone implants and wireless orthopedic devices. Here, we will introduce the frontier of exploring biological responses on implant materials, especially nanoscale materials and wireless devices, and how the design and fabrication of implant surfaces based on these explorations can ultimately aid in bone health. But first, we must discuss what biological events would need to be controlled by wireless orthopedic medical devices to treat bone problems.

3.2   Fundamentals of the Interface Between Sensors and Bone

3.2.1   Events at the Sensor–Bone Tissue Interface

Bone is a very complicated biological system that consists of both hierarchical structures and living bone remodeling units. The architecture of bone is composed of nanofibrous collagen matrices, noncollagenous proteins, and nanocrystalline calcium phosphate (mainly hydroxyapatite). Bone remodeling units involve three major types of bone cells: osteoblasts (bone forming cells), osteocytes (bone-maintaining cells), and osteoclasts (bone-resorbing cells). Therefore, one can expect the interactions at the interface of bone tissue and implant materials are very intricate events, as illustrated in Fig. 2.

Basically, all of these events can be categorized as host responses toward the implant and, conversely, material responses to the host (Puleo and Nanci 1999). In the present context, it is important to realize that an implanted wireless sensor will interact in all of the events in Fig. 2. It is also important to mention that for this application, an optimal sensor will not only sense new bone growth (and whether it is occurring), but it will also improve new bone growth. Though all of these events described in Fig. 1 are of great importance when designing a wireless medical device, there are several body responses that are of particular interest since such information can be used to determine bone health next to the sensor. (1) Protein adsorption is the immediate event once a material is implanted. Also, the type and density of the adsorbed proteins can regulate cell adhesion and subsequent cellular activities. (2) Adhesion of osteogenic cells, subsequent bone deposition, and bone remodeling are crucial factors at a successful orthopedic sensor–bone interface. (3) Sensor surface properties are also important events closely related to biomaterial cytocompatibility, immune or inflammatory responses that may ultimately cause implant failure. Immune and inflammatory cell responses are not introduced

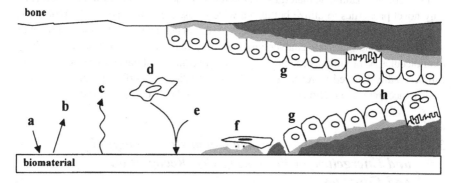

Fig. 2 Events at the bone–implant interface. (**a**) Protein adsorption from blood and tissue fluids, (**b**) protein desorption, (**c**) surface changes and material release, (**d**) inflammatory and connective tissue cells approach the implant, (**e**) possible targeted release of matrix proteins and selected adsorption of proteins, (**f**) formation of lamina limitans and adhesion of osteogenic cells, (**g**) bone deposition on both the exposed bone and implant surfaces, and (**h**) remodeling of newly formed bone (adapted from Puleo and Nanci (1999))

in this chapter; however, the reader is cautioned to bear in mind that they are of exceptional importance at the bone–implant interface because if the sensor is encapsulated in scar tissue, it will not be able to "sense."

Understandably, the bone–sensor interface is difficult to characterize because of the complexities of the various biological responses and in vivo environment. Therefore, in vitro bone cell culture models become practical and effective tools to initially investigate biological responses and cell functions on sensor surfaces. Most of the experimental results discussed in this chapter are, thus, based on bone cell culture models. Importantly, nanotechnology (or more specifically, nanomaterials) has been shown to control such events and, thus, should be an integral part of an implantable wireless sensor.

3.2.2 Novel Properties of Nanomaterials/Nanotechnology

Nanoscale materials are defined as materials (e.g., particles, fibers, tubes, etc.) with size scales within 1–100 nm in at least one dimension. The research and development aimed at understanding and working with (e.g., detecting, measuring and manipulating) these kinds of materials has been called nanotechnology (Balasundaram 2007). Nanotechnology emerged in the last century and has shown extraordinary potential in biomedical research applications. This potential originates from unique properties of nanomaterials compared to bulk conventional (e.g., micron grain size) materials, such as: (1) more surface reactivity as a result of much larger surface areas; (2) greatly enhanced mechanical properties (such as high ductility and high yield strength) due to various mechanisms such as increased grain boundary sliding and short-range diffusion-healing; (3) exceptional magnetic, optical, and electrical properties because of stacking, alignment, and orientation of nanoscale building blocks (grains, supermolecules, etc.); and (4) homogeneity and high purity in composition or structure thanks to reacting or mixing at the molecular or atomic level.

Furthermore, nanoscale materials or structures provide the bone community with not only novel properties to utilize but also a wide landscape to understand the biological responses on a material surface. One important reason behind this is that the comparable size of nanoscale materials enables researchers to really detect, interact, and analyze biomolecules or bio-microstructures. The other reason is that nanostructured materials can be readily tailored to reveal extraordinary variations in surface properties, which leads to accurate observations of their effects on cellular or tissue responses.

3.3 The Role of Sensor Surface Chemistry, Topography, and Energetics on Promoting Cell Recognition and Function

The surface properties of sensors that cells and tissues recognize (through initial protein interactions) in vivo or in vitro are chemistry, topography, and energy. As surface chemistry and topography both contribute to surface energy, which is directly

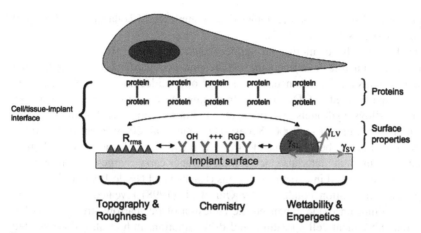

Rrms:root mean square roughness; γ_{SV}: interfacial tension between solid and vapor; γ_{SL}: interfacial tension between solid and liquid; γ_{LV}: interfacial tension between liquid and vapor; RGD:arginine-glycine-aspartic acid

Fig. 3 Interactions between cells and sensor surface properties at the implant–cell interface

related to surface wettability, many researchers have been using surface energy to characterize the interface between materials and cells. Figure 3 summarizes the sensor (or implant)–cell interface and interactions between cells, proteins, and sensor surface properties. It is worth remembering that proteins always dictate the interactions between surface properties and cells in an aqueous environment (body fluids, culture media, etc.). Therefore, understanding the impact of surface properties on protein adsorption and activity can also be an effective tool to explore biological responses on different sensor surfaces. In this section, we follow these rationales to explain the roles surface chemistry, topography, and energetics play on cellular functions, emphasizing how nanotechnology is promoting such interactions.

3.3.1   Surface Chemistry

There are many factors of surface chemistry that affect biological responses on sensors, such as the inherent chemistry (i.e., the chemical or phase composition and crystallinity of materials), possibly conjugated functional groups/molecules on the surface, and surface charge/polarity due to the configuration of such chemical groups. Different types of implant materials have been shown to be either bio-inert or bioactive, which results in either morphological fixation or bioactive fixation to the host tissue (Cao and Hench 1996). The different states of fixation indicate that the inherent chemistry of materials alters the responses of biological systems, and this information has guided researchers to seek a variety of implant materials. For example, the materials for orthopedic implants range from metals (CoCrMo alloys, titanium (Ti) and its alloys, and stainless steel), ceramics (alumina, zirconia, titania, and hydroxyapatite), ultra high molecular weight polymers (polyethylene, polyurethane, and poly-lactic-co-glycolic acid [PLGA]) to biologically synthesized substances (such as mineralized complexes of collagens, calcium, and phosphate) (Balasundaram 2007).

On the other hand, modifying surfaces with functional groups have demonstrated a crucial role in influencing cell responses regardless of the underlying inherent chemistry. For instance, a wide range of peptides (most notably, arginine–glycine–aspartic acid (RGD)) have been conjugated on various material surfaces (Ti and its alloys, hydrogels, polymers, etc.) to improve bone cell functions in vitro and in vivo (Schuler et al. 2006; Kroese-Deutman et al. 2005; Picart et al. 2005). In addition, it is also believed that electrostatic interactions play a role in biological responses to implant materials since cell membranes carry charges and virtually all interfaces are charged in aqueous solutions (Haynes and Norde 1994; Wilson et al. 2005). An observation conducted by Qiu et al. (1998) revealed that positively charged indium tin oxide enhanced the adhesion of rat marrow stromal cells but impaired subsequent cell spreading and differentiation. Itoh et al. (2006) created electrically polarized hydroxyapatite with pores and reported increased bone growth and decreased osteoclast activity. They attributed these effects to the electrical polarity on surfaces with pores.

Clearly, one aspect of how nanotechnology has been used to improve bone growth has been through altered surface chemistry. Nanotechnology exhibits a great potential to improve implant efficacy by manipulating the discrete surface chemistry regions of nanomaterials. There are several reasons for this:

1. Nanomaterials can provide much larger surface areas, more substructures (such as grain boundaries), and more active surfaces for chemical modification. For instance, nano-fibrous poly(L-lactic acid) scaffolds modified by entrapping a large amount of gelatin molecules on the scaffold surface demonstrated improvements in osteoblast adhesion, proliferation, and compressive modulus (Liu et al. 2006).
2. A variety of novel nanomaterial shapes (dots, rods, tubes, cages, etc.) can be used for patterned, anisotropic, or multiple functionalizations. One example is that incorporating segments of growth factors and antibiotics into anodized nanotubular Ti can further enhance new bone formation and inhibit infection (Yao and Webster 2006).
3. Controllable assembly of nanomaterials can be realized or adjusted to certain chemical effects by a bottom-up strategy that establishes structures from tiny building blocks (e.g., atoms, molecules, etc.). For example, layer-by-layer nano-assembly of poly(lysine)/alginate above a gelatin (or extracellular matrix) surface resulted in a 200-fold decrease in the adhesion of human fibroblasts (cells that contribute to soft, not hard, granulation tissue formation) compared to the untreated surfaces (Ai et al. 2003).

3.3.2 Topography and Roughness

Another predominant surface property that can be easily modified through nanotechnology to build better interfaces between sensors and bone is surface topography.

Among several ways of describing surface topography, roughness is a major parameter; its effects on bone tissue/cell responses has been studied intensely during the last two decades (Thomas and Cook 1985). Biologically inspired by the hierarchical micron-to-nanostructure of bones, which osteoblasts are naturally accustomed to in the body, an approach of creating nanometer roughness or nanoscale features on orthopedic implant surfaces can increase the efficacy of many orthopedic implant chemistries. This rationale has been applied to a variety of nanoscale materials (including metals (Webster and Ejiofor 2004), carbon nanofibers/nanotubes (Price et al. 2003), polymers (Washburn et al. 2004), ceramics (Webster et al. 2000), and polymer/ceramic composites (Webster and Smith 2005)), and increased osteoblast functions (such as adhesion, proliferation, differentiation, and calcium deposition, etc.) on these nanostructured materials have been reported. Meanwhile, a general positive correlation between nanoscale roughness and biological responses (i.e., increased roughness corresponds to enhanced osteoblast functions) has been established (Anselme et al. 2000; Linez-Bataillon et al. 2002). One explanation of the nanometer roughness-enhanced osteoblast functions is that large surface areas associated with increased roughness can adsorb more proteins (fibronectin, vitronectin, etc.) important for osteoblast adhesion. However, other studies suggested that proteins adsorb differentially with variations in nanoscale surface roughness (Wilson et al. 2005), perhaps relating nanometer surface features to changes in surface energetics important for controlling selective protein adsorption.

Several other nanoscale topographical features that cannot be described simply by roughness may also mediate bone cells responses at the bone–sensor interface. Texture and alignment of surface patterns have been reported to influence both osteoblast functions and orientation on materials. For example, Biggs et al. (2007) reported a decrease in osteoblast adhesion and spreading on highly ordered nanopits (120 nm in diameter and 100 nm in depth) than randomly distributed ones, noticing that surface roughness on both substrates was similar because the number of pits per area on each was approximately the same. Zhu et al. (2005) created nanoscale polystyrene grooves with spacings of 150 nm and a depth of 60–70 nm and observed alignment of osteoblast-like cells and cell-produced collagen fibers along nanogrooves. Recently, thickness gradients of coatings in the nanometer regime have also been found to influence bone cell responses. For instance, a poly(3-octylthiophene-2,5-diyl) film with thickness gradients (120–200 nm) showed a higher proliferation ratio compared to tissue culture polystyrene controls after 1 day of incubation, while cell adhesion after a period of 4 h was not affected by changes in thickness over 10–60 nm (Rincon et al. 2009).

3.3.3 Wettability and Surfaces Energetics

To further understand the effects of nanoscale surface topography and roughness on osteoblast responses, researchers have defined surface energy. Surface energy is closely related to wettability (hydrophobicity or hydrophilicity) of a surface, and the basic relationship is given by Young's equation:

$$\cos\theta = (\gamma_{SV} - \gamma_{SL}) / \gamma_{LV},$$

where θ is the contact angle, γ_{SV} is the interfacial tension (or surface energy) between the solid and vapor, γ_{SL} is the interfacial tension between the solid and liquid, and γ_{LV} is the interfacial tension between the liquid and vapor.

It has been widely observed that hydrophilic surfaces are energetically favorable for the adhesion and subsequent activities of osteoblasts as well as many other types of cells, probably relating to the increased adsorption of several hydrophilic proteins than hydrophobic proteins less important for cell adhesion. Measuring surface energy demonstrates an association of the relatively high surface energy on hydrophilic surfaces compared to hydrophobic ones, and the criterion of ca. $\theta=65°$ that differentiates between the two regimes has been suggested (Vogler 1999; Lim et al. 2008). One clear study created both hydrophilic (high surface energy) and hydrophobic (low surface energy) surfaces by modifying chemistry on quartz surfaces and showed that hydrophilic surfaces induced homogeneously-spaced osteoblastic cell growth and mineral deposition, and enhanced the quantity (e.g., area) and quality (e.g., mineral-to-matrix ratio) of mineralization by osteoblasts (Lim et al. 2008). The authors suggested that surface energy effects on osteoblast differentiation, especially mineralization, could be correlated with surface energy dependent changes in spatial cell growth. Roughness also alters surface energy and resultant wettability, which establishes critical connections between topography and surface energetics. Although it has not been well substantiated, a trend of increased surface energy resulting from the increased presence of nanometer surface features enhances osteoblast functions has been shown in a number of studies (Anselme 2000; Das et al. 2007; Khang et al. 2008).

In summary, surface energy is closely related to surface wettability and can be modified through both surface chemistry and roughness. An idea of integrating surface chemistry, topography (roughness), and wettability into a unified expression of surface energetics and correlating such properties with biological responses will be of great importance to the understanding and mediation of protein, cell, and tissue responses on nanotechnology-created sensor surfaces.

3.4 Novel Sensor Surfaces: Better Biological Responses and Better Performance

Although the reasons for enhanced osteoblast functions on nanostructured materials are under intense investigation, the evidence that osteoblasts perform better on nanostructured surfaces is overwhelming. As one example, osteoblast responses to carbon nanotubes (CNTs) have been studied. Due to their excellent electrical conductivity, great mechanical strength, and unique chemical–biological properties (Iijima 1991; Lin et al. 2004; Webster et al. 2004; Smart et al. 2006; Zanello 2006), many studies have shown that CNTs are promising for bone sensor applications.

CNTs are macromolecules of carbon, classified as single walled carbon nanotubes (SWCNTs), diameter 0.4–2 nm, and multiwalled carbon nanotubes (MWCNTs), diameter 2–100 nm. In theoretical and experimental results, CNTs have an electric-current-carrying capacity 1,000 times higher than copper wires (Avouris et al. 2003). Thus, CNTs have been considered to improve the electrical properties of such sensors as well. For instance, CNT-TiN nanocomposites, composed of 12% CNTs by volume, exhibited a 45% increased electrical conductivity over TiN materials (Jiang and Gao 2005). Since bone regenerates under electrical conduction, many researchers have used CNTs to promote bone growth. For example, nano-composites of polylactic acid and MWCNTs have been shown to increase osteo-blast proliferation by 46% and calcium production by greater than 300% when an alternating current was applied to the substrate in vitro (Supronowicz et al. 2002).

Also, CNTs have been used to increase the mechanical strength of nanocompos-ite scaffolds for bone tissue engineering. SWCNTs, functionalized with phosphates and poly(aminobenzene sulfonic acid), can substitute as collagen to direct the anisotropic crystallization of hydroxyapatite (HA) reaching a thickness of 3 mm after 14 days of mineralization, resulting in composites that can be used as supporting scaffolds for orthopedic applications (Zhao et al. 2005). MWCNTs improve the mechanical properties of the as-aligned HA composite coatings (Chen et al. 2007). Balani et al. (2007) showed that a MWCNT reinforced HA coating promoted human osteoblast proliferation in vitro compared to a normal HA coating, and osteoblasts were observed near MWCNTs regions.

CNTs exhibit a large surface area to volume ratio, which may support and promote cell attachment. It has been investigated that a large number of human osteoblasts adhered more on carbon nanofiber (CNF) micro-patterns than polycarbonate urethane (Khang et al. 2006). The morphology of osteoblasts extended in all directions within CNT scaffolds formed on polycarbonate membranes (Aoki et al. 2005). Nanophase poly(lactic-co-glycolic) acid (PLGA) casts of CNF (average diameter=60 nm) com-pacts possessed a higher degree of nanometer surface roughness by approximately 50%, which increased osteoblast adhesion after 1 h compared with PLGA casts of conventional CNFs (average diameter=200 nm) (Price et al. 2004).

Moreover, osteoblast-like cells also significantly enhanced their adhesion on vertically aligned MWCNTs arrays on substrates (another great sensor), which were prepared by lithography. It was observed that the periodicity and alignment of vertically aligned MWCNTs considerably influenced the growth, shape, and orien-tation of the osteoblasts (Giannona et al. 2007). Zanello also showed that single/multi-walled CNTs scaffolds with/without surface modifications are suitable for osteosarcoma ROS 17/2.8 cell proliferation. Lastly, the cell densities and transform-ing growth factor-$\beta 1$ secreted by human osteoblast-like (Saos2) cells were higher on MWCNT scaffolds compared with polystryrene and polycarbonate scaffolds.

Importantly, in the studies mentioned above, CNTs were synthesized by a number of different techniques, yet all showed greater osteoblast functions. CNTs can be produced by a laser furnace, the arc, and chemical vapor deposition (CVD). However, CVD is a scalable and controllable method to obtain high purity CNTs. Sato et al. (2005) revealed that cobalt (Co) particles could extend their catalytic ability by combining with Ti particles when growing MWCNTs by CVD and forming a strong contact between Ti and MWCNTs. All of these techniques speak well for creating a sensor composed of electrically active CNTs that can also positively interact with bone cells.

Importantly, for such sensor design, cell responses transduce and transmit a variety of chemical and physical signals to produce specific substances and proteins within specific tissues and organs. The in situ sensing of proteins from specific cell induction processes can be employed as a signal of specific cellular responses or bone regeneration. For example, in a previous study, MWCNTs grown by the CVD technique out of anodized Ti nanotubes (Fig. 4) exhibited excellent electrochemical properties and stability. An in vitro study showed the redox of proteins could be enhanced on MWCNTs grown from an anodized nanotubular Ti (MWCNT-Ti) electrode in order to sense bone growth (Sirivisoot et al. 2007; Sirivisoot and Webster 2008), and such a sensor can promote osteoblast proliferation and differentiation after 21 days on MWCNT-Ti. MWCNT-Ti can thus be used as a bio-sensing material to generate electrical signals, which can be interpreted later as information.

Moreover, coupling drug delivery to implantable wireless sensors enables on-command diffusion-controlled drug delivery systems by using radio-frequency, which is a new approach in the orthopedic field. There is a need for new technology for the delivery of soluble/insoluble or stable/unstable therapeutic compounds locally to improve the bioavailability of drugs. Polypyrrole is a conductive electro-active polymer, which has been shown to be biocompatible and has been proposed

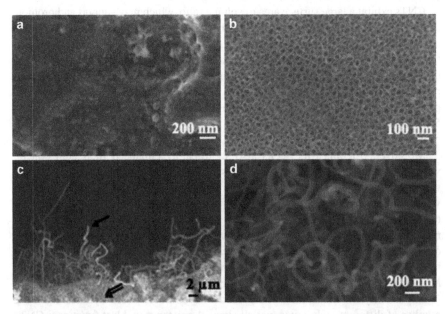

Fig. 4 SEM micrographs of multiwalled carbon nanotubes (MWCNTs) grown out of an anodized nanotubular surface: (**a**) currently implanted conventional titanium (Ti) which has no sensing capabilities; (**b**) anodized nanotubular Ti which has no sensing abilities; and (**c**) side and (**d**) top views of MWCNT grown out of anodized nanotubular Ti which has bone growth sensing abilities

for several in vivo applications, such as a conductor for the electrical stimulation of cells. PPy can be synthesized on Ti for local drug delivery, promoting bone regeneration, and carrying therapeutic drugs. Typical doses of antibiotic/anti-inflammatory drugs need to be large for hip/knee implants. Improvement in the delivery efficiency and localization may also result by controlling the thickness and the area of the polypyrrole membrane. Thus, sensors are being developed that can sense new bone formation, and if that is not happening, release drugs to promote new bone growth. Similar attempts are being attempted for decreasing orthopedic implant infection and inflammation.

In summary, this exciting new nanomaterial, or CNTs, may become a powerful tool in bone sensor applications in terms of their nanofeatures, superior mechanical properties, and higher electrical conductivity to sense and control cellular behavior. Continuing the careful studies on CNTs may discover additional potentials for this novel biomaterial for engineering biocompatible nanomaterials and nanodevices for various sensor applications.

4   Summary and Remaining Challenges for Wireless Monitoring and Sensing Medical Devices

Wireless physiological measurement systems, like other innovations, seek to reduce risks (Hao and Foster 2008). However, all new technologies have unanswered questions. The major remaining questions concerning all wireless medical devices (whether implantable or not) are as follows (Hao and Foster 2008):

- Reliability – the main challenge is to make sure that information reliably gets to its destination. The reliability of a wireless physiological measurement system relies on many aspects, such as reliable wireless communication between nodes, efficient computation in each sensor node, and stable software programming (Hongliang et al. 2006).
- Biocompatibility – the shape, size, and materials are restricted for sensors that directly act on the human body. One solution is to package the sensor nodes in biocompatible materials and use nanomaterials, which appear to optimally interact with biological tissues (Hongliang et al. 2006).
- Portability – the size of the sensors used in wireless physiological measurement systems and in implantable sensors needs to be small and lightweight.
- Privacy and security – there are big security issues to be considered, such as eavesdropping, identity spoofing (i.e., the assumption of a trusted user's security credentials during a communication session), and redirection of private data to unauthorized persons. Security can be improved using data encryption. It is necessary to protect private data from improper access and alteration.
- Lightweight protocols for wireless communication – must support self-organizing networks (including security aspects) and be able to perform data collection and routing.

- Energy-aware communication – it is desirable for sensors to transmit at low power. An energy-aware protocol is necessary to allow sensors to negotiate their transmission power to a minimum.
- RF radiation safety – the electromagnetic radiation must be within the recommended SAR limits. In the United States, the FCC has set the safe exposure limit to a SAR level at or below 1.6 W kg^{-1} in 1 g of tissue. In Europe, the European Union Council has adopted the SAR limit of 2 W kg^{-1} in 10 g of tissue.

However, in light of these high demands for sensors, the excitement concerning developing more efficient and effective medical devices based on wireless medical sensor technology makes this journey worth every effort.

References

Ai H, Jones SA and Lvov YM. Biomedical applications of electrostatic layer-by-layer nano-assembly of polymers, enzymes, and nanoparticles. Cell Biochemistry and Biophysics 2003; 29: 23–43

American Academy of Orthopedic Surgeons (AAOS) (2010). Available at: http://www.aaos.org/Research/stats/patientstats.asp. Accessed November 10, 2010

Anselme K. Osteoblast adhesion on biomaterials. Biomaterials 2000; 21: 667–681

Anselme K, Bigerelle M, Noel B, Dufresne E, Judas D, Iost A and Hardouin P. Qualitative and quantitative study of human osteoblast adhesion on materials with various surface roughnesses. Journal of Biomedical Materials Research 2000; 49: 155

Aoki N, Yokoyama A, Nodasaka Y, Akasaka T, Uo M, Sato Y, et al. Cell culture on a carbon nanotube scaffold. Journal of Biomedical Nanotechnology 2005; 1: 402–404

Avouris P, Appenzeller J, Martel R and Wind SJ. Carbon nanotube electronics. Proceedings of the IEEE 2003; 91(11): 1772–1784

Balani K, Anderson R, Laha T, Andara M, Tercero J, Crumpler E, et al. Plasma-sprayed carbon nanotube reinforced hydroxyapatite coatings and their interaction with human osteoblasts in vitro. Biomaterials 2007; 28(4): 618–624

Balasundaram G (2007) Nanomaterials for better orthopedics. In: Webster TJ, editor. Nanotechnology for the Regeneration of Hard and Soft Tissues. Singapore: World Scientific, pp 53–78

Biggs MJP, Richards RG, Gadegaard N, Wilkinson CDW and Dalby MJ. The effects of nanoscale pits on primary human osteoblast adhesion formation and cellular spreading. Journal of Materials Science. Materials in Medicine 2007; 18: 399–404

Bren L. Joint replacement: an inside look. FDA Consumer 2004; 38(2). Available at: http://www.fda.gov/fdac/features/2004/204_joints.html. Accessed November 10, 2010

Cao WP and Hench LL. Bioactive materials. Ceramics International 1996; 22: 493–507

Chen Y, Zhang TH, Gan CH and Yu G. Wear studies of hydroxyapatite composite coating reinforced by carbon nanotubes. Carbon 2007; 45(5): 998–1004

Das K, Bose S and Bandyopadhyay A. Surface modifications and cell–materials interactions with anodized Ti. Acta Biomaterialia 2007; 3: 573–585

Fass L (2007) Patient centric healthcare. 3rd IET Int. Conf. on Medical Electrical Devices & Technology (London, 2–3 October 2007)

Feied C, Jordan N, Kanhouwa M and Kavanagh J (2006) The new world of healthcare work: a Microsoft white paper. UK Focus International Lecture, The Royal Academy of Engineering (7 November 2007)

Giannona S, Firkowska I, Rojas-Chapana J and Giersig M. Vertically aligned carbon nanotubes as cytocompatible material for enhanced adhesion and proliferation of osteoblast-like cells. Journal of Nanoscience and Nanotechnology 2007; 7: 1679–1683 (1675)

Hao and Fostee RY. (2008) Wireless body sensor networks for health-monitoring applications. Physiol meas. 2008; 29(11): R27–56

Haynes CA and Norde W. Globular proteins at solid/liquid interfaces. Colloids Surfaces B Biointerfaces 1994; 2: 517

Hongliang R, Meng MQH and Chen X (2006) Physiological information acquisition through wireless biomedical sensor networks. *Proc. 2005 IEEE Int. Conf. on Information Acquisition (Hong Kong and Macau, China, 27 June–3 July 2005)*

Iijima S. Helical microtubules of graphitic carbon. Nature 1991; 354(6348): 56–58

Itoh S, Nakamura S, Nakamura M, Shinomiya K and Yamashita K. Enhanced bone ingrowth into hydroxyapatite with interconnected pores by electrical polarization. Biomaterials 2006; 27: 5572–5579

Jiang L and Gao L. Carbon nanotubes-metal nitride composites: a new class of nanocomposites with enhanced electrical properties. Journal of Materials Chemistry 2005; 15(2): 260–266

Jones VM, et al. (2006) Remote monitoring for healthcare and for safety in extreme environments. In: *M-Health: Emerging Mobile Health Systems.* Istepanian R., Caxmin-Arayan, S., Pattichis C. (eds) Berlin: Springer, pp 561–574. ISBN 0387265589

Khang D, Sato M, Price RL, Ribbe AE and Webster TJ. Selective adhesion and mineral deposition by osteoblasts on carbon nanofiber patterns. International Journal of Nanomedicine 2006; 1(1): 65–72

Khang D, Lu J, Yao C, Haberstroh KM and Webster TJ. The role of nanometer and sub-micron surface features on vascular and bone cell adhesion on titanium. Biomaterials 2008; 29: 970–983

Kroese-Deutman HC, Van Den Dolder J, Spauwen PHM and Jansen JA. Influence of RGD-loaded titanium implants on bone formation in vivo. Tissue Engineering 2005; 11: 1867–1875

Lim JY, Shaughnessy MC, Zhou Z, Noh H, Vogler EA and Donahue HJ. Surface energy effects on osteoblast spatial growth and mineralization. Biomaterials 2008; 29: 1776–1784

Lin Y, Taylor S, Li H, Fernando KAS, Qu L, Wang W, et al. Advances toward bioapplications of carbon nanotubes. Journal of Materials Chemistry 2004; 14(4): 527–541

Linez-Bataillon P, Monchau F, Bigerelle M and Hildebrand HF. In vitro MC3T3 osteoblast adhesion with respect to surface roughness of Ti6Al4V substrates. Biomolecular Engineering 2002; 19: 133

Liu XH, Won YJ and Ma PX. Porogen-induced surface modification of nano-fibrous poly(L-lactic acid) scaffolds for tissue engineering. Biomaterials 2006; 27: 3980–3987

Mendelson Y, Duckworth RJ and Comtois G (2006) A wearable reflectance pulse oximeter for remote physiological monitoring. *Proc. 28th IEEE EMBS Ann. Int. Conf. (New York City, USA, 30 August–3 September 2006).* Piscataway, NJ: Institute of Electrical and Electronics Engineers, vol 1, pp 912–915

Pattichis C, Kyriacou E, Voskarides S, Pattichis M, Istepanian R and Schizas C. Wireless tele-medicine systems: an overview. IEEE Antennas & Propagation Magazine 2002; 44: 143–153

Penders J, et al. (2008) Human++: from technology to emerging health monitoring concepts. *Int. Workshop on Wearable and Implantable Body Sensor Networks (BSN 2008) (1–3 June 2008)* pp 94–98. Accessed November 10, 2010

Picart C, Elkaim R, Richert L, Audoin T, Arntz Y, Cardoso MD, et al. Primary cell adhesion on RGD-functionalized and covalently crosslinked thin polyelectrolyte multilayer films. Advanced Functional Materials 2005; 15: 83–94

PillCam® Capsule Endoscopy. Available at: http://www.givenimaging.com/. Accessed November 10, 2010

Price RL, Waid MC, Haberstroh KM and Webster TJ. Selective bone cell adhesion on formulations containing carbon nanofibers. Biomaterials 2003; 24: 1877–1887

Price RL, Ellison K, Haberstroh KM and Webster TJ. Nanometer surface roughness increases select osteoblast adhesion on carbon nanofiber compacts. Journal of biomedical materials research Part A 2004; 70(1): 129–138

ProeT EX Project Webpage. Available at: http://www.proetex.org/. Accessed November 10, 2010

Puleo DA and Nanci A. Understanding and controlling the bone-implant interface. Biomaterials 1999; 20(23:24): 2311–2321

Qian H, Loizou PC and Dorman MF. A phone-assistive device based on bluetooth technology for cochlear implant users. IEEE Transactions on Neural Systems and Rehabilitation Engineering 2003; 11: 282–287. Accessed November 10, 2010

Qiu Q, Sayer M, Kawaja M, Shen X and Davies JE. Attachment, morphology, and protein expression of rat marrow stromal cells cultured on charged substrate surfaces. Journal of Biomedical Materials Research 1998; 42: 117

Rincon C, Chattopadhyay S and Meredith C (2009) Development of semi-conductor biomaterials for regulating cell growth. *Annual Meeting of AICHE (Salt Lake City)*

Roman R, Lopez J and Gritzalis S. Situation awareness mechanisms for wireless sensor networks. IEEE Communications Magazine 2008; 46: 102–107. Accessed November 10, 2010

Sato S, Kawabata A, Kondo D, Nihei M and Awano Y. Carbon nanotube growth from titanium, cobalt bimetallic particles as a catalyst. Chemical Physics Letters 2005; 402(1–3): 149–154

Saxby R (2007) How silicon will transform healthcare. *3rd IET Int. Conf. on Medical Electrical Devices & Technology (London, 2–3 October 2007).* Accessed November 10, 2010

Schofield I and Heath H (1999) *Healthy Ageing: Nursing Older People.* Amsterdam: Elsevier Health Sciences. Accessed November 10, 2010

Schuler M, Owen GR, Hamilton DW, De Wilde M, Textor M, Brunette DM, et al. Biomimetic modification of titanium dental implant model surfaces using the RGDSP-peptide sequence: a cell morphology study. Biomaterials 2006; 27: 4003–4015

Schwiebert L, Gupta SKS and Weinmann J (2001) Research challenges in wireless networks of biomedical sensors. *MobiCom '01: Proc. 7th Ann. Int. Conf. on Mobile Computing and Networking (New York, USA)* pp 151–165. Accessed November 10, 2010

Seyedi A and Sikdar B (2008) Modeling and analysis of energy harvesting nodes in body sensor networks. *Int. Workshop on Wearable and Implantable Body Sensor Networks (BSN 2008) (1–3 June 2008)* pp 175–178. Accessed November 10, 2010

Shin H, Jo S and Mikos AG. Modulation of marrow stromal osteoblast adhesion on biomimetic oligo[poly(ethylene glycol) fumarate] hydrogels modified with Arg-Gly-Asp peptides and a poly(ethylene glycol) spacer. Journal of Biomedical Materials Research 2002; 61: 169–179. Accessed November 10, 2010

Sirivisoot S and Webster TJ. Multiwalled carbon nanotubes enhance electrochemical properties of titanium to determine in situ bone formation. Nanotechnology 2008; 19(29): 295101

Sirivisoot S, Yao C, Xiao X, Sheldon BW and Webster TJ. Greater osteoblast functions on multi-walled carbon nanotubes grown from anodized nanotubular titanium for orthopedic applications. Nanotechnology 2007; 18(36): 365102

Smart SK, Cassady AI, Lu GQ and Martin DJ. The biocompatibility of carbon nanotubes. Toxicology of Carbon Nanomaterials 2006; 44(6): 1034–1047

Smith R (2004) *Bone Health and Osteoporosis: A Report of the Surgeon General.* Rockville, MD: U.S. Department of Health and Human Service, Public Health Service, Office of the Surgeon General, pp 68–700

Spectrums for Body Sensor Networks: GE Healthcare Monitoring Solution (2007) Presentations to FCC, ET Docket No. 06-135 (24 July 2007). Accessed November 10, 2010

Supronowicz PR, Ajayan PM, Ullmann KR, Arulanandam BP, Metzger DW and Bizios R. Novel current-conducting composite substrates for exposing osteoblasts to alternating current stimulation. Journal of Biomedical Materials Research 2002; 59(3): 499–506

Texas Instruments (2007) CC2420 2.4 GHz IEEE 802.15.4/ZigBee-Ready RF Transceiver: Datasheet. Accessed November 10, 2010

Thomas KA and Cook SD. An evaluation of variables influencing implant fixation by direct bone apposition. Journal of Biomedical Materials Research 1985; 19: 875

TinyOS Community Website. Available at: http://www.tinyos.net. Accessed November 10, 2010

Townsend KA, Haslett JW, Tsang TKK, El-Gamal MN and Iniewski K (2005) Recent advances and future trends in low power wireless systems for medical applications. *Proc. 9th Int. Database Engineering & Application Symposium (IDEAS'05).* Piscataway, NJ: Institute of Electrical and Electronics Engineers Computer Society

Vogler EA. Water and the acute biological response to surfaces. Journal of Biomaterials Science. Polymer Edition 1999; 10: 1015–1045

Walter P, Kisv'arday ZF, G"ortz M, Alteheld N, Rossler G, Stieglitz T and Eysel UT. Cortical activation via an implanted wireless retinal prosthesis. *Investigative Ophthalmology & Visual Science* 2005; 46: 1780–1785. Accessed November 10, 2010

Washburn NR, Yamada KM, Imon Jr CG, Kennedy SB and Amis EJ. High-throughput investigation of osteoblast response to polymer crystallinity: influence of the nanometer-scale roughness on proliferation. Biomaterials 2004; 25: 1215–1223

Webster TJ. Nanophase ceramics as improved bone tissue engineering materials. American Ceramic Society Bulletin 2003; 82: 23–28

Webster TJ and Ejiofor JU. Increased osteoblast adhesion on nanostructured metals: Ti, Ti_6Al_4V, and CoCrMo. Biomaterials 2004; 19: 4731–4739

Webster TJ and Smith TA. Increased osteoblast function on PLGA composites containing nanophase titania. Journal of Biomedical Materials Research. Part A 2005; 74(4): 677–686

Webster TJ, Ergun C, Doremus RH, Siegel RW and Bizios R. Enhanced osteoblast functions on nanophase ceramics. Journal of Biomedical Materials Research 2000; 51: 475–479

Webster TJ, Waid MC, McKenzie JL, Price RL and Ejiofor JU. Nano-biotechnology: carbon nanofibres as improved neural and orthopaedic implants. Nanotechnology 2004; 15(1): 48–54

Wilson CJ, Clegg RE, Leavesley DI and Pearcy MJ. Mediation of biomaterial–cell interactions by adsorbed proteins: a review. Tissue Engineering 2005; 11: 1–18

Wong ACW, Kathiresan G, Chan CKT, Eljamaly O and Burdett A (2007) A 1V wireless transceiver for an ultralow-power wireless SoC for biomedical applications. *Proc. IEEE Eur. Conf. Solid State Circuits (ESSIRC) (September 2007)*. Accessed November 10, 2010

Xiao Y. Accountability for wireless LANs, ad hoc networks and wireless mesh networks. IEEE Communications Magazine 2008; 46: 116–126. Accessed November 10, 2010

Yao C and Webster TJ. Anodization: a promising nano-modification technique of titanium implants for orthopedic applications. Journal of Nanoscience and Nanotechnology 2006; 6: 2682–2692

Yeatman EM (2006) Rotating and gyroscopic MEMS energy scavenging. *Int. Workshop on Wearable and Implantable Body Sensor Networks (BSN 2006) (3–5 April 2006)* p 4. Accessed November 10, 2010

Yoo H-J, Song S-J, Cho N and Kim H-J (2007) Low energy on-body communication for BSN. *Body Sensor Networks (Aachen, Germany, 26–28 March 2007)* pp 15–20. Accessed November 10, 2010

Zanello LP. Electrical properties of osteoblasts cultured on carbon nanotubes. Micro & Nano Letters 2006; 1(1): 19–22

Zhao B, Hu H, Mandal SK and Haddon RC. A bone mimic based on the self-assembly of hydroxyapatite on chemically functionalized single-walled carbon nanotubes. Chemistry of Materials 2005; 17(12): 3235–3241

Zhao Y, Hao Y, Alomainy A and Parini C. UWB on-body radio channel modelling using ray theory and sub-band FDTD. IEEE Transactions on Microwave Theory and Techniques 2006; 54: 1827–1835. Accessed November 10, 2010

Zhou G, Lu J, Wan C-Y, Jarvis MD and Stankovic JA (2008) BodyQoS: adaptive and radio-agnostic QoS for body sensor networks. *27th IEEE Conf. on Computer Communications (13–18 April 2008)* pp 565–573. Accessed November 10, 2010

Zhu B, Lu Q, Yin J, Hu J and Wang Z. Alignment of osteoblast-like cells and cell-produced collagen matrix induced by nanogrooves. Tissue Engineering 2005; 11: 825–834. Accessed November 10, 2010

Zhu Y, Keoh SL, Sloman M, Lupu E, Dulay N and Pryce N (2008) A policy system to support adaptability and security on body sensors. *Int. Workshop Wearable and Implantable Body Sensor Networks (BSN 2008) (1–3 June 2008)* pp 37–40. Accessed November 10, 2010

ZigBee Alliance Website. Available at: http://www.zigbee.org/en/index.asp. Accessed November 10, 2010

Zimmerman TG (1996) Personal area networks: near-field intrabody communication. IBM Systems Journal 1996; 35: 609–617. Accessed November 10, 2010

Chapter 3
Monitoring Inflammation and Infection via Implanted Nanosensors

Batur Ercan, Nhiem Tran, and Thomas J. Webster

Abstract Sensors are being used for a wide range of applications. Some of the most intriguing applications are those that involve detecting infection and inflammation. This chapter covers some fundamentals of medical device infection as well as inflammation and how sensors are being used to detect such events. It is hoped that such sensors would detect bacteria presence before infection sets in and then indicate to a clinician that such infection has occurred. It is also hoped that sensors will be able to determine early events in inflammation to avoid chronic inflammation that has plagued numerous medical devices. This chapter summarizes such advances.

Keywords Inflammation • Infection • Nanotechnology • Sensors • Drug delivery

1 Inflammation

1.1 Introduction

When an implant is inserted, some biological reactions take place near the implanted area. This is the body's response to this newly implanted foreign material. A successful orthopedic implant promotes the adhesion of osteoblasts on the implant surface and the formation of new bone tissue, incorporating the implant into the body. However, if the body encapsulates the implant by the formation of soft fibrous tissue and tries to separate it as much as possible from surrounding bone, this is the sign of an unsuccessful implantation. The overall aim for numerous medical devices is to minimize the amount of fibrous tissue formation around the implant and to maximize appropriate new tissue growth.

T.J. Webster (✉)
School of Engineering, Brown University, 182 Hope Street, Providence,
RI 02917, USA
e-mail: thomas_webster@brown.edu

T.J. Webster (ed.), *Nanotechnology Enabled In situ Sensors for Monitoring Health*,
DOI 10.1007/978-1-4419-7291-0_3, © Springer Science+Business Media, LLC 2011

1.2 Host Response to Foreign Materials

After the implantation procedure, the body follows a sequence of local events during the healing response. These are acute inflammation, chronic inflammation, granulation tissue formation, foreign body reaction, and fibrosis. The steps and predominant cell types present during these steps of wound healing are shown in Fig. 1 (Karlsson et al. 2004).

Inflammation is the reaction of vascularized living tissue to local injury. It activates a series of events which may heal tissue after implantation by recruiting parenchymal cells or cells intended to encapsulate the implant with fibrous tissue. The initial inflammatory response is activated regardless of the type of biomaterial and the location of injury. Acute inflammation is a short-term (few days) response to the injury. Immediately after the surgery, the blood flow to the injury site increases. This implies an increased flow of proteins (some of them can promote anchorage-dependent cell attachment), nutrition, immune cells, cells that can help recovery, mesenchymal stem cells, etc. The first cells to appear at the implantation site are white blood cells, mainly neutrophils. Afterwards, neutrophils recruit monocytes to the inflammation area (where monocytes will further differentiate into macrophages which are well known to aid in implant material degradation). Neutrophils are then recruited to the site of inflammation by chemical mediators (chemotaxis) to phagocytose microorganisms and foreign materials (Anderson et al. 1996). They attach on the surfaces of the biomaterial by adsorbed proteins, basically immunoglobulin G (IgG) and complement-activated fragments (such as C3b) through a process called "opsonization." Due to the relative size difference between biomaterials and phagocytotic cells, cells clearly cannot internalize the implant. This leads to a condition called "frustrated phagocytosis," which is the activation of phagocytic cells to produce extracellular products that attempt to degrade the biomaterial and, at the same time, recruit more cells to the implant site. Macrophages then secrete degradative

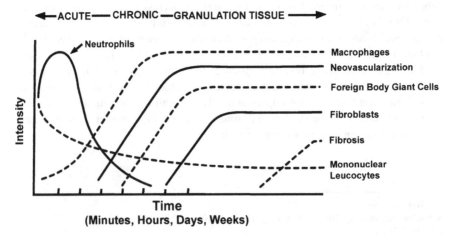

Fig. 1 Time vs. intensity graph for the wound healing responses showing the predominant cell types at the inflammation site. (Adapted and redrawn from (Karlsson et al. 2004))

agents (such as superoxides and free radicals) which severely damage both the juxtaposed tissue and possibly the implant.

A persistent inflammatory response leads to chronic inflammation (Anderson et al. 1996). The main cell types observed during chronic inflammation are monocytes, macrophages, and lymphocytes. Neovascularization (formation of new blood vessels) also starts during this step of wound healing. Macrophages are the most important type of cells in chronic inflammation due to the secretion of a great number of biologically active products such as: proteases, arachidonic acid metabolites, reactive oxygen metabolites, coagulation factors, and growth factors (which are important to recruit and promote fibroblast functions).

The third step in the foreign body response is granulation tissue formation. Fibroblasts form granulation tissue. This tissue is the hallmark of the healing response. It is granular in appearance and contains many small blood vessels (Maki and Tambyah 2001). In addition, macrophages fuse together to form foreign body giant cells to attempt to phagocytose the foreign materials much more effectively. The amount of granulation tissue determines the extent of fibrosis.

The foreign body reaction, the fourth step in wound healing, contains foreign body giant cells and granulation tissue (which includes fibroblasts, capillaries, macrophages, etc.). Biomaterial surface properties are the key determining factor in this step. It has been shown that smooth surfaces (micron smooth) induce a foreign body reaction which is composed of macrophages one to two cells in thickness (Anderson et al. 1996). However, as the roughness (micron rough) increases, so does the foreign body reaction.

The last step in the wound healing response is fibrosis, which is the fibrous tissue encapsulation of the implant. The successfulness of implantation generally depends on the proliferation capacity of the cells in the tissue. Tissues containing labile (proliferate throughout time) or stable (expanding) cells are less likely to have entered the fibrosis step. However, if the cells comprising that tissue are stable (not growing, limited proliferation capacity), the chances of fibrous tissue formation are relatively higher. As stated before, fibrous encapsulation is not desired in bone tissue engineering applications because it leads to failure of the implant due to stress–strain imbalances. Quite simply, fibrous tissue cannot support physical stresses that some tissues can (like bone or cartilage). The desired situation for an orthopedic implant is the recruitment of parenchymal cells, around the implant as soon as possible.

2 Infection

2.1 Introduction

Another problem that plagues medical devices is infection, particularly catheters. Catheters are widely used in hospitals for a number of applications including liquid drainage/injection and instrument access. One of the leading problems with catheters is bacterial infection. Bacteria infect up to 54% of all catheters (Maki and Tambyah 2001; Trerotola 2000) and cause many serious complications including patient death. For example, catheter infection is associated with a mortality rate of 12–25% among critically ill patients (Sanders et al. 2008). Catheter-associated urinary tract infection (CAUTI) is the most common type (accounting for 40%) of hospital-acquired infections ("nosocomial infection"), resulting in serious complications such as bloodstream infection, and even death (Sanders et al. 2008). Each year, in the U.S. acute-care hospitals and extended-care facilities, CAUTI affects approximately one million patients, who then have increased institutional death rates (Trerotola 2000). Chronic indwelling urinary catheters also increase the risk of infection, accounting for 80% of all nosocomial urinary tract infections (Roe et al. 2008).

Infections have also been reported to be the most severe complication of tunneled dialysis catheters which are used by approximately 20% of hemodialysis patients in the U.S., resulting in serious systemic infections, including endocarditis, osteomyelitis, epidural abscess, septic arthritis, and even death (Maki and Tambyah 2001). Significantly, 14% of the deaths in people undergoing dialysis in 1996 were due to infection (Maki and Tambyah 2001).

Another example of serious catheter and catheter-related infections concerns the use of central venous catheters. Infection may occur in 3–7% of the approximately three million central venous catheters inserted annually in the U.S., resulting in 150,000–250,000 nosocomial bacteremias (bloodstream infection). The mortality rate for this type of infection is from 10 to 20% (Leitman and Valavanur 1999). Importantly, the average cost of care is $45,000 per patient with this type of infection (Pronovost et al. 2006).

Lastly, ventilator-associated pneumonia (VAP), an infection which can occur in the lung in patients undergoing mechanical ventilation through the use of endotracheal tubes, is the most common hospital-acquired infection in intensive care units. VAP occurs in approximately 9–18% of patients being intubated (Kollef et al. 2008). Bacteria colonize the endotracheal tubes and then quickly spread to the lungs due to breathing patterns of the patients. VAP usually occurs within 10 days after endotracheal intubation, resulting in an increased length of hospital stay and associated health-care costs (average cost is $40,000 per patient) and significant mortality (Kollef et al. 2008).

2.2   The Biofilm

Catheter infections are difficult to treat because of biofilm formation. After catheterization, bacteria (from the patient's own skin, hospital personnel, or equipment) quickly attach and some adhere irreversibly to the catheter surface, secreting polymeric-like substances (composed of mostly polysaccharides) and form a biofilm which consists of bacteria and a polymeric-like matrix (Fig. 2). Bacteria in a biofilm can escape from the film and enter the blood, lungs, etc., causing serious problems. Biofilms tenaciously bind to today's catheter surfaces. More importantly, bacteria in biofilms are extremely resistant to antibiotic treatment due to the slow

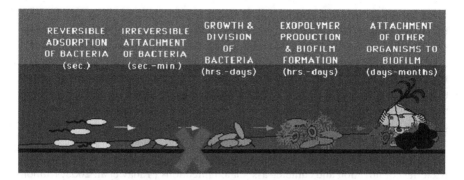

| REVERSIBLE ADSORPTION OF BACTERIA (sec.) | IRREVERSIBLE ATTACHMENT OF BACTERIA (sec.-min.) | GROWTH & DIVISION OF BACTERIA (hrs.-days) | EXOPOLYMER PRODUCTION & BIOFILM FORMATION (hrs.-days) | ATTACHMENT OF OTHER ORGANISMS TO BIOFILM (days-months) |

Fig. 2 Formation of biofilm by bacteria (figure adapted from (http://bioinfo.bact.wisc.edu))

transport of antibiotic molecules through the polymeric-like biofilm substance, altered microenvironment within the biofilm, and higher numbers of persistent cells (i.e., cells resistant to many types of stress) within the biofilm (compared to plank-tonic [free-floating] cells) (Donlan 2001; Davies 2003). Among the most common pathogens found in infected catheters and endotracheal tubes are *Staphylococcus epidermidis* (24%), *Staphylococcus aureus* (20%), and *Pseudomonas aeruginosa* (25%) (Donlan 2001; Chastre and Fagon 2002).

2.3 Current Methods of Preventing and Treating Catheter and Endotracheal Tube Infections

Conventional, traditional techniques to prevent catheter infections (such as practicing good aseptic techniques, systemic administration of antibiotics, etc.) have not provided satisfactory results to date. For example, systemic antibiotic (vancomycin and gentamicin) administration alone without catheter removal is only 22–37% effective in treating tunneled catheter-associated bacteremia (Maki and Tambyah 2001). The reasons for the ineffectiveness of these traditional prac-tices are the resistance of bacteria toward antibiotics. Placing an antibiotic solution into the catheter lumen is another method to prevent and treat catheter infections. However, this method has a problem of leading to the development of antibiotic resistant bacteria. Therefore, this method is generally not recommended by the Kidney Disease Outcomes Quality Initiative and the Centers for Disease Control and Prevention (Yahav et al. 2008).

With the number of bacteria that are resistant to antibiotic treatment alarmingly increasing, attention has been recently paid to coating or impregnating catheters with nonantibiotic antimicrobial agents (Yahav et al. 2008). Since catheters them-selves are an important source of infection, and bacterial adhesion to the catheter surface is important in the pathogenesis of infection, the most promising and straightforward strategy toward decreasing infection is to fabricate catheters that resist bacteria attachment (Berra et al. 2008). However, researchers are still looking for such antibacterial coating materials that can be easily applied and are inexpensive. For example, silver and silver compounds (such as chlorhexidine-silver sulfadiazine), one of the most studied materials for antibacterial coatings on catheters or for impregnating in catheters, have yielded mixed results toward preventing infections. Some studies as early as the 1950s and later in the 1970s, and 1990s, showed that silver coatings could reduce the incidence of catheter-associated infections (Winson 1997; Akiyama and Okamoto 1979; Liedberg and Lundeberg 1990). In contrast, a large number of recent studies have demonstrated that catheters coated or impregnated with silver or silver compounds did not reduce the incidence of bacterial colonization and catheter-associated infections (Yahav et al. 2008; Kumon et al. 2001; Trerotola et al. 1998; Logghe et al. 1997) or lacked efficacy (Trerotola 2000; Dunser et al. 2005). Especially, in some studies, silver-coated catheters were shown to be ineffective against *S. aureus, S. epidermidis, Enterobacter aerogenes, Klebsiella pneumoniae, P. aeruginosa, Candida albicans*, and *Escherichia coli* (Bach et al. 1999). There is even concern about the increase in *staphylococcal* infections in male patients using silver oxide impregnated catheters (Bach et al. 1999; Gaonkar and Modak 2003). Coupled to this controversy of using silver towards decreasing catheter infections is their high cost (many hospitals have opted not to use silver-coated catheters due to expense, and in those few hospitals that do, only high-risk patients are receiving them). Endotracheal tubes coated with silver can cost as much as $100 (whereas noncoated endotracheal tubes are a couple of dollars).

Other examples of the ineffectiveness of current antibacterial catheters are catheters impregnated with nitrofurazone which is an antibiotic. Such catheters were shown to be ineffective against vancomycin-resistant *enterococci* and were predicted to be ineffective against gram-negative strains (Lawrence and Turner 2005). In addition, most of the antimicrobial catheters do not prevent infection of long-term (more than 2 weeks) catheterization (Trerotola 2000; Kumon et al. 2001).

Recently, some other materials, such as those which release nitric oxide (NO), have been investigated as antibacterial coatings for catheters. NO, a diatomic free radical, has been shown to have antibacterial properties against some gram-negative and gram-positive bacteria (such as *P. aeruginosa, S. aureus*, and *S. epidermidis*) (Ahearn et al. 2000; Nablo et al. 2005). Therefore, some research has focused on incorporating NO donor molecules (diazeniumdiolate) into polymers to make antibacterial coatings. However, such approaches require complex fabrication methods which drive up costs. For example, the most common method is incorporating diamine-containing organosilanes into a sol–gel matrix, then exposing the sol–gel to high pressure of NO gas so that the diamine coordinates two NO molecules to form a NO donor molecule.

Besides NO-releasing coatings, TiO_2 (titanium dioxide) has also been studied for antibacterial coatings (Evans and Sheel 2007; Page et al. 2007). However, this approach is less applicable to coating catheters since it is difficult to coat TiO_2 onto polymers because of adhesion problems between the oxide coating and the polymer substrates (Woodyard et al. 1996; Jeom Sik et al. 2005; Girshevitz et al. 2008; Navarro-Alarcon and Cabrera-Vique 2008).

3 Examples of Biosensors Used for Infection and Inflammation

3.1 Ex Vivo

Of course, detection of bacteria both inside and outside the body is important. In a recent study, a new biological application of quantitative Raman spectroscopy was proposed to detect low levels of bacteria and inflammation (Bergholt and Hassig 2009). Native human plasma C-reactive protein (CRP) was used as a clinical biomarker of bacterial infection and tissue damage. This protein circulates in the blood and the concentration rises as inflammation occurs. For the first time, the Raman spectrum of CRP in a buffered aqueous solution was acquired using 785 nm excitation. The concentration of CRP was measured in blood plasma, using near-infrared (NIR) Raman spectroscopy. Spectra were acquired with an in situ Inspector Raman spectrometer using 785 nm excitation. Raman spectra were collected from blood plasma drawn from 40 individuals. Quantitative predictions of CRP were made by means of partial least squares (PLS) analysis and a variable selection method Interval PLS (IPLS). The similarity of the features in the PLS regression vector to that of CRP's Raman spectrum illustrated that the prediction was sensitive to the molecular information determined by the Raman scattered light (Bergholt and Hassig 2009). The IPLS algorithm was applied to optimize the calibration model to near clinical accuracy. This study provided critical information toward the feasibility of using Raman spectroscopy for quantitative measurements of CRP in blood plasma, and, thus, to determine infection and inflammation (Fig. 3) (Bergholt and Hassig 2009).

3.2 In Situ

Of course, for a number of years, researchers have created drug-release devices that respond to the in situ external environment to release drugs. Although such devices do not provide information from inside the body to outside the body, they are a sort of internal sensor. Clearly, today's efforts are more focused on developing sensors that can transmit information internally to externally, although some of these earlier drug-release devices are intriguing for anti-infection and anti-inflammatory applications. For example, several studies have aimed at constructing novel-triggered drug delivery systems that release antimicrobials at specific locations at required times (Wu and Grainger 2006). Such systems can be triggered by certain endogenous host infection responses such as inflammation-related enzymes, thrombin activity, or microbial proteases. Drug-conjugated polymers synthesized using 1,6-hexane diisocyanate (HDI), polycaprolactone diol (PCL), and the fluoroquinolone antibiotic (ciprofloxacin) polymerized into the polymer backbone release drug as the polymer degrades by an inflammatory cell-derived enzyme, cholesterol esterase (Woo et al. 2000).

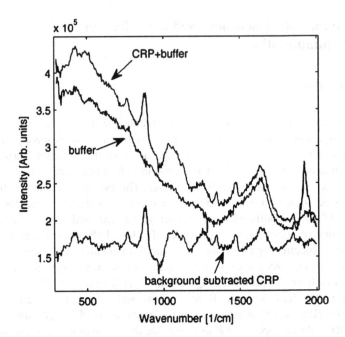

Fig. 3 Raman spectrum of CRP in a buffered aqueous solution and the pure buffer using 785 nm excitation. Also shown is the background-subtracted Raman spectrum of CRP. Adapted from (Navarro-Alarcon and Cabrera-Vique 2008)

Microbiological assessment showed that released ciprofloxacin possessed antimicrobial activity against *P. aeruginosa* after 10 days. Combinations of different enzymatically labile polymer–drug linkages and modification of the degradable polymer chemistry may solve this problem (Harten et al. 2005).

Importantly, significant increases in thrombin-like activity have been reported in *S. aureus*-infected wounds (Tanihara et al. 1999). Researchers, thus, developed a novel peptide to link an insoluble polymer matrix with antimicrobials specifically cleaved by thrombin (Tanihara et al. 1998). A PVA–peptide–gentamicin conjugate was developed and investigated both in vitro and in vivo. Released gentamicin amounts were dependent on local thrombin concentration associated with *S. aureus* infection with bacteriocidal effects observed in animal models of *S. aureus* infection, thus providing an intelligent system for fighting infection. Again, however, information concerning bacteria presence is not relayed to outside the body as would be desirable for potential clinical intervention.

In contrast, there are a number of techniques that exist which can determine implant infection, but do nothing to simultaneously remove it. Specifically, one technique involves using impedance spectroscopy as a powerful method of analyzing the complex electrical resistance of a system (Lisdat and Schafer 2008). Like all electrochemical biosensors, impedimetric sensors are bioelectronic devices that make use of the interactions of biomolecules with a conductive (or semiconducting) transducer surface (Rickert et al. 1996). The detection process involves the formation of a recognition complex between the sensing biomolecule and the analyte at the interface of the electronic transducer, which directly or indirectly alters the electrical properties of the recognition surface. The various efforts directed at specific and sensitive detection can be classified according to the electrode material (metals, metal oxides, glassy carbon, semiconductors, etc.), the electrode geometry (conventional electrode arrangement or interdigitated electrodes), the analyte (proteins, antibodies, nucleic acids, etc.), or according to the amplification protocol used (label-free, enzyme labels, conducting polymer films, nanoparticles, etc.) (Lisdat and Schafer 2008). It is sensitive to surface phenomena and changes in bulk properties (Lisdat and Schafer 2008). In the field of biosensors, it is particularly well suited for the detection of binding events on the transducer surface (Lisdat and Schafer 2008). First examples of its use were reported at the end of the 1980s; however, this method has found increasing applications in recent years due to advances made in instrumentation. Besides the detection of biorecognition processes, it is a valuable tool for characterizing surface modifications, such as those that occur during the immobilization of biomolecules on the transducer (Lisdat and Schafer 2008).

As already implied, impedimetric analyte detection is frequently applied for immunosensing (Fig. 4) (Lisdat and Schafer 2008). One source of impetus for the development of sensors with impedimetric detection are the demands of point-of-care diagnostics, which require rather simple equipment and the potential for label-free analysis (Lisdat and Schafer 2008). Two immobilization strategies are possible for immunosensing. The sensor chip can either be modified with antibodies that bind the respective antigen molecule, or the antigen itself can be immobilized,

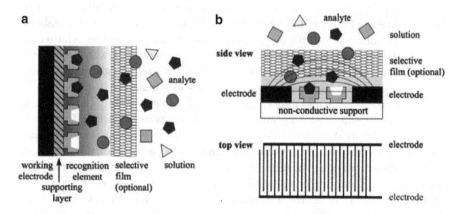

Fig. 4 Schematic of an impedimetric biosensor using a modified working electrode as a sensor (**a**) or an in-plane arrangement of two electrodes and (**b**) immobilization of the biocomponent occurs on the electrode in (**a**) and between and/or on the electrodes in (**b**). Adapted from (Lisdat and Schafer 2008)

which then binds the complementary antibody (Lisdat and Schafer 2008). In both cases, the binding event results in a change in the electrical surface properties, although larger changes can be expected in the latter case because of the high molecular weight and the low dielectric constant of the antibodies.

Initially, capacitive sensors were preferably used for the detection of immune reactions (Rickert et al. 1996; Taira et al. 1993; Berggren et al. 1998; Bataillard et al. 1988). Charge transfer can be diminished by using long-chain self-assembled monolayers (SAM) for example. When used in combination with a potential-jump method, such sensors can yield very sensitive measurements (Berggren et al. 1998; Berggren and Johansson 1997). It has been also shown that a continuous binding analysis is possible, allowing a mathematical description of the binding behavior to be obtained (Bataillard et al. 1988).

Resistance-based sensors have been also developed (Hleli et al. 2006; Jie et al. 1999; Pak et al. 2001). For example, the human mammary tumor-associated glycoprotein was detected by specific antibodies immobilized on gold by spontaneous adsorption (Lisdat and Schafer 2008). The binding of the complementary antigen resulted in a change in the charge transfer resistance (Jie et al. 1999). The resistance-based detection was also used to study receptor–ligand interactions and allowed the detection of, for example, odorant molecules (Hou et al. 2007). Another system for immune analysis used ultrathin platinum layers and evaluated the conductivity changes based on an impedance model analysis (Pak et al. 2001). However, often the sensitivity of direct binding detection is insufficient to meet practical demands (Lisdat and Schafer 2008). Thus, several amplification strategies have been developed, as will be discussed in the next section.

Further recent developments are cell-based assays for the detection of specific analytes or for the screening of large libraries of substances for their biological effects. This relies on the concept of effect monitoring (Lisdat and Schafer 2008). Here, parameters are needed that report the response of the whole biological system with respect to the stimulus under investigation. Impedance is one of these parameters because it can be used for the indirect detection of metabolic activity, cell adhesion on surfaces, response to potential drugs, and for cytotoxicity tests (Lisdat and Schafer 2008). Studies have been reported in which the status of cell cultures or even single cells have been analyzed (Lisdat and Schafer 2008). For example, apoptosis-induced changes in human colon cancer HT-29 cell shape have been investigated (Lisdat and Schafer 2008). Impedance is not only suitable for obtaining information on the status of surface-fixed cells but can also be used to detect the number of cells in solution. This has been intensively investigated in microbiology and has been used for the detection, quantification, and identification of bacteria, as well as in micromachined devices for cell counting or cell differentiation (Lisdat and Schafer 2008). For the specific capture of bacterial cells, antibodies or bacteriophages can be used (Lisdat and Schafer 2008). Impedimetric analysis has also been applied to the detection of immune cells.

4 Conclusions

In summary, while there have been numerous efforts for ex vivo diagnosis of inflammation and infection (from blood samples or other bodily fluids), much effort is needed to develop in vivo sensors. Some cues can be taken from drug delivery systems which have been shown to take external information to release drugs to fight infection or reduce inflammation; however, these systems do not transfer information from inside to outside of the body. Thus, the future is bright for developing sensors that can combine existing technologies to sense inflammation and infection surrounding implants.

References

Ahearn, D.G., et al., *Effects of hydrogel/silver coatings on in vitro adhesion to catheters of bacteria associated with urinary tract infections.* Curr Microbiol, 2000. **41**(2): p. 120–5.

Akiyama, H. and S. Okamoto, *Prophylaxis of indwelling urethral catheter infection: clinical experience with a modified Foley catheter and drainage system.* J Urol, 1979. **121**(1): p. 40–2.

Anderson, J.M., Gristina A.G., Hanson S.R., Harker L.A., Johnson R.J., Merritt K., Naylor P.T., Schoen F.J., "Host reactions to biomaterials and their evaluation", In: Ratner B.D, Hoffman A.S, Schoen A.S, Lemons J.E, editors. "Biomaterials science: an introduction to materials in medicine", San Diego: Academic, 165–214, (1996)

Bach, A., et al., *Efficacy of silver-coating central venous catheters in reducing bacterial colonization.* Crit Care Med, 1999. **27**(3): p. 515–21.

Bataillard, P., F. Gardies, N. Jaffrezicrenault, C. Martelet, B. Colin, and B. Mandrand, *Direct detection of immunospecies by capacitance measurements.* Anal Chem, 1988. **60**: p. 2374.

Berggren, C. and G. Johansson, *Capacitance Measurements of Antibody–Antigen Interactions in a Flow System.* Anal Chem, 1997. **69**: p. 3651.

Berggren, C., B. Bjarnason, G. Johansson, *An immunological Interleukine-6 capacitive biosensor using perturbation with a potentiostatic step.* Biosens Bioelectron, 1998. **13**: p. 1061.

Bergholt, M.S. and S. Hassig, *Quantification of C-reactive protein in human blood plasma using near infrared Raman spectroscopy.* Analyst, 2009. **134**: p. 2123–7.

Berra, L., et al., *Antimicrobial-coated endotracheal tubes: an experimental study.* Intensive Care Med, 2008. **34**(6): p. 1020–9.

Chastre, J. and J.-Y. Fagon, *Ventilator-associated Pneumonia.* Am J Respir Crit Care Med, 2002. **165**(7): p. 867–903.

Davies, D., *Understanding biofilm resistance to antibacterial agents.* Nat Rev Drug Discov, 2003. **2**(2): p. 114–22.

Donlan, R.M., *Biofilms and device-associated infections.* Emerg Infect Dis, 2001. **7**(2): p. 277–81.

Dunser, M.W., et al., *Central venous catheter colonization in critically ill patients: a prospective, randomized, controlled study comparing standard with two antiseptic-impregnated catheters.* Anesth Analg, 2005. **101**(6): p. 1778–84.

Evans, P. and D.W. Sheel, *Photoactive and antibacterial TiO_2 thin films on stainless steel.* Surf Coatings Technol, 2007. **201**(22–23): p. 9319–24.

Gaonkar, T.A. and S.M. Modak, *Comparison of microbial adherence to antiseptic and antibiotic central venous catheters using a novel agar subcutaneous infection model.* J Antimicrob Chemother, 2003. **52**(3): p. 389–96.

Girshevitz, O., Y. Nitzan, and C.N. Sukenik, *Solution-deposited amorphous titanium dioxide on silicone rubber: a conformal, crack-free antibacterial coating.* Chem Mater, 2008. **20**(4): p. 1390–6.

Harten, R.D., D.J. Svach, R. Schmeltzer, and K.E. Uhrich, *Salicylic acid-derived poly(anhydride-esters) inhibit bone resorption and formation in vivo.* J Biomed Mater Res A, 2005. **72A**(4): p. 354–62.

Hleli, S., C. Martelet, A. Abdelghani, N. Burais, and N. Jaffrezic-Renault, *Atrazine analysis using an impedimetric immunosensor based on mixed biotinylated self-assembled monolayer.* Sens Actuators B, 2006. **113**: p. 711.

Hou, Y.X., N. Jaffrezic-Renault, C. Martelet, A.D. Zhang, J. Minic-Vidic, T. Gorojankina, M.A. Persuy, E. Pajot-Augy, R. Salesse, V. Akimov, L. Reggiani, C. Penetta, E. Alfinito, O. Ruiz, G. Gomilla, J. Samitier, and A. Errachid, *A novel detection strategy for odorant molecules based on controlled bioengineering of rat olfactory receptor I7.* Biosens Bioelectron, 2007. **22**: p. 1550.

http://bioinfo.bact.wisc.edu, accessed August 2010.

Jeom Sik, S., et al., *Surface modification of silicone rubber by ion beam assisted deposition (IBAD) for improved biocompatibility.* J Appl Polym Sci, 2005. **96**(4): p. 1095–101.

Jie, M., C.Y. Ming, D. Jing, L.S. Cheng, L.H. Na, F. Jun, and C.Y. Xiang, *An electrochemical impedance immunoanalytical method for detecting immunological interaction of human mammary tumor associated glycoprotein and its monoclonal antibody.* Electrochem Commun, 1999. **1**: p. 425.

Karlsson M., "Nanoporous alumina, a potential bone implant coating", Uppsala University, PhD Thesis (2004).

Kollef, M.H., et al., *Silver-coated endotracheal tubes and incidence of ventilator-associated pneumonia: the NASCENT randomized trial.* JAMA, 2008. **300**(7): p. 805–13.

Kumon, H., et al., *Catheter-associated urinary tract infections: impact of catheter materials on their management.* Int J Antimicrob Agents, 2001. **17**(4): p. 311–6.

Lawrence, E.L. and I.G. Turner, *Materials for urinary catheters: a review of their history and development in the UK.* Med Eng Phys, 2005. **27**(6): p. 443–53.

Leitman, I.M. and A.S. Valavanur, *Central venous catheters: can we control the infections?* Curr Surg, 1999. **56**(1): p. 24–9.

Liedberg, H. and T. Lundeberg, *Silver alloy coated catheters reduce catheter-associated bacteriuria.* Br J Urol, 1990. **65**(4): p. 379–81.

Lisdat, F. and D. Schafer, *The use of electrochemical impedance spectroscopy for biosensing.* Anal Bioanal Chem, 2008. **391**: p. 1555–67.

Logghe, C., et al., *Evaluation of chlorhexidine and silver-sulfadiazine impregnated central venous catheters for the prevention of bloodstream infection in leukaemic patients: a randomized controlled trial.* J Hosp Infect, 1997. **37**(2): p. 145–56.

Maki, D.G. and P.A. Tambyah, *Engineering out the risk for infection with urinary catheters.* Emerg Infect Dis, 2001. **7**(2): p. 342–7.

Nablo, B.J., A.R. Rothrock, and M.H. Schoenfisch, *Nitric oxide-releasing sol-gels as antibacterial coatings for orthopedic implants.* Biomaterials, 2005. **26**(8): p. 917–24.

Navarro-Alarcon, M. and C. Cabrera-Vique, *Selenium in food and the human body: a review.* Sci Total Environ, 2008. **400**(1-3): p. 115–41.

Page, K., et al., *Titania and silver-titania composite films on glass-potent antimicrobial coatings.* J Mater Chem, 2007. **17**(1): p. 95–104.

Pak, S.C., W. Penrose, and P.J. Hesketh, *An ultrathin platinum film sensor to measure biomolecular binding.* Biosens Bioelectron, 2001. **16**: p. 371.

Pronovost, P., et al., *An intervention to decrease catheter-related bloodstream infections in the ICU.* N Engl J Med, 2006. **355**(26): p. 2725–32.

Rickert, R., W. Gopel, W. Beck, G. Jung, and P. Heiduschka, *A 'mixed' self-assembled monolayer for an impedimetric immunosensor.* Biosens Bioelectron, 1996. **11**: p. 757.

Roe, D., et al., *Antimicrobial surface functionalization of plastic catheters by silver nanoparticles.* J Antimicrob Chemother, 2008. **61**(4): p. 869–76.

Sanders, J., et al., *A prospective double-blind randomized trial comparing intraluminal ethanol with heparinized saline for the prevention of catheter-associated bloodstream infection in immunosuppressed haematology patients.* J Antimicrob Chemother, 2008. **62**(4): p. 809–15.

Taira, H., K. Nakano, M. Maeda, and M. Takagi, *Electrode Modification by Long-Chain, Dialkyl Disulfide Reagent Having Terminal Dinitrophenyl Group and Its Application to Impedimetric Immunosensors.* Anal Sci, 1993. **9**: p. 199.

Tanihara, M., Y. Suzuki, Y. Nishimura, K. Suzuki, Y. Kakimaru, *Thrombin-sensitive peptide linkers for biological signal-responsive drug release systems.* Peptides (New York), 1998. **19**(3): p. 421–5.

Tanihara, M., Y. Suzuki, Y. Nishimura, K. Suzuki, Y. Kakimaru, and Y. Fukunishi, *A novel microbial infection-responsive drug release system.* J Pharm Sci, 1999. **88**(5): p. 510–4.

Trerotola, S.O., *Hemodialysis catheter placement and management.* Radiology, 2000. **215**(3): p. 651–8.

Trerotola, S.O., et al., *Tunneled hemodialysis catheters: use of a silver-coated catheter for prevention of infection – a randomized study.* Radiology, 1998. **207**(2): p. 491–6.

Winson, L., *Catheterization: a need for improved patient management.* Br J Nurs, 1997. **6**(21): p. 1229–32; 1234; 1251–2.

Woo, G.L.Y., M.W. Mittelman, and J.P. Santerre, *Synthesis and characterization of a novel biodegradable antimicrobial polymer.* Biomaterials, 2000. **21**(12): p. 1235–46.

Woodyard, L.L., et al., *A comparison of the effects of several silver-treated intravenous catheters on the survival of staphylococci in suspension and their adhesion to the catheter surface.* J Control Release, 1996. **40**(1–2): p. 23–30.

Wu, P. and D. Grainger, *Drug device combinations for local drug therapies and infection prophylaxis,* Biomaterials, 2006. **27**: p. 2450–67.

Yahav, D., et al., *Antimicrobial lock solutions for the prevention of infections associated with intravascular catheters in patients undergoing hemodialysis: systematic review and meta-analysis of randomized, controlled trials.* Clin Infect Dis, 2008. **47**(1): p. 83–93.

Chapter 4
DNA-Based Nanotechnology Biosensors for Surgical Diagnosis

Yupeng Chen and Hongchuan Yu

Abstract This chapter covers the use of DNA in biological sensing devices. Although DNA has been used for a number of years for in vitro sensing applications, there are several unique challenges to using DNA as in vivo sensors. This chapter covers some of the more significant advances in this field highlighting how these challenges are being met. In doing so, it emphasizes various ways DNA can be immobilized to implants to sense implant functionality. It also emphasizes various forms of DNA highlighting the advantages and disadvantages for in vivo sensor applications.

Keywords Sensors • DNA • Nanotechnology • In vivo • Immobilization

1 Introduction and Background

A typical biosensor design includes three main parts: target recognition elements, signal transduction elements, and signal amplification elements (not always required) (McNaught et al. 1997). Among these three components, the target recognition element is the key to contact, detect, and recognize useful medical signals for further analysis and diagnosis. More importantly, using functional biomaterials as recognition elements enhances the sensitivity, selectivity, stability, and lifetime of a biosensor. Today, biological engineering has created a variety of biomaterials for sensor applications. Such materials can be either natural biological components (such as tissues, microorganisms, organelles, cell receptors, enzymes, antibodies, and nucleic acids) (Xia et al. 2010; McLaughlin et al. 2010; Guo et al. 2010) or biologically derived materials (so-called biomimetic materials) (Lin et al. 2010;

Y. Chen (✉)
Department of Orthopaedics, School of Engineering, Brown University, 184 Hope Street, Providence, RI 02917, USA
e-mail: Yupeng_Chen@Brown.edu

Huang et al. 2010; Bi et al. 2006). For example, deoxyribonucleic acid (DNA) or DNA-derived materials are receiving intensive attention for biosensor applications.

Because DNA itself is highly selective in base-pairing interactions between complementary sequences and DNA contains the genetic instructions used in the development and functioning of all known living organisms, they are valuable for diagnosis and are relatively easy to be specifically detected. As an example, single-stranded DNA was immobilized on a biosensor probe, where base-pairing interactions were recognized and hybridized with a target DNA to the surface to release signals, and such signals were recorded and analyzed for diagnosis, as shown in Fig. 1. A signal can be reported optically, mechanically, or electrochemically. For instance, in an optical method, gold nanoparticles modified with single-stranded DNA change color during the hybridization of the target DNA and immobilized DNA (Storhoff et al. 1998). Another widely used method involves electrochemical detection. Because electrochemical reactions (such as oxidation and reduction) contain an electron transition process, such processes result in electrical signals. Such signals can be detected directly to save an expensive signal transduction system and increase the sensitivity of diagnosis. There are two detection mechanisms: one is hybridization-based DNA detection as mentioned above and the other is enzymatic-based DNA detection. For the enzymatic-based DNA detection, DNA-related enzymes are introduced into the biological recognition system and changes in the amount of these enzymes correlate to specific biological processes (such as the deletion or fusion of the target DNA). For example, when a sensor experiences a specific process, the enzyme level either increases or decreases, resulting in amplification or reduction of the signal. The enzymatic process is highly specific to a DNA sequence, which makes it ideal for DNA mismatch detection, as shown in Fig. 2 (Wei et al. 2010).

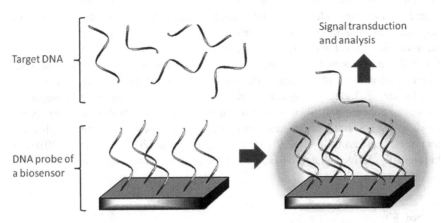

Fig. 1 A schematic DNA nanotechnology biosensor design. Target DNA is captured at the DNA probe and the resulting signal is transduced for analysis (adapted and redrawn from Drummond et al. (2003))

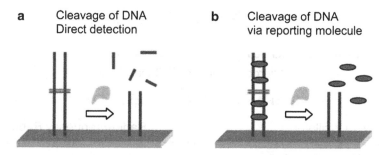

Fig. 2 Electrochemically based DNA nanotechnology sensors for protein analytes. (**a**) Direct detection after specific DNA enzymatic process. (**b**) Detection of extralabeled reporter after specific DNA enzymatic process (adapted and redrawn from Wei et al. (2010))

Traditionally, DNA-based biosensors do not interact with the biological system during the detection process and are known as "passive" sensors. They are usually used to detect the target molecules in samples collected from host biological systems. However, in surgical applications, it is often difficult to collect samples from inspected organs or tissues. Especially, surgeons sometimes need to monitor the target organ or tissue to diagnose diseases or evaluate the efficacy of a surgery. In this manner, the "passive" DNA-based sensors cannot satisfy the need of surgical diagnosis. Therefore, a novel emerging technique, in situ biosensors, has been developed for various surgical applications (Leegsma-Vogt et al. 2004; Wilkins and Atanasov 1996). For instance, a biosensor technique, called electromyography, has been used to analyze and diagnose muscle or nerve functions inside body conditions (Hoch and Zieve 2008). Thin electrodes are placed in soft tissues to help analyze and record electrical activity in the muscles. Another field where biosensors are widely used in surgical diagnosis is orthopedics. A telemetric device for orthopedic implants has been invented for "real time" diagnosis. Typically, the telemetry system with a small implantable transmitter uses wireless communication with an external computer system. The data are transmitted at high frequency or radio-frequency pulses, as shown in Fig. 3 (Durr et al. 1999; Graichen and Bergmann 1991; Graichen et al. 1999). Basically, the semiconductor strain gauges are placed at the inner circumference of the neck of an orthopedic implant cavity. A small loop antenna outside the implant (the pacemaker is fed through and forms a single loop antenna outside the Ti at one end of the implant) is used to transmit the radio-frequency pulses to the external receiving antenna. The radio-frequency receiver of the external device (TELEPORT, as shown in Fig. 3) is connected to a microcontroller to analyze the measured voltages. Finally, computer software interprets the data sent from the microcontroller to display the real-time load components. In addition, researchers have also focused on using DNA-based biosensors to rapidly detect tumor cells or cancer markers in the blood or tissues (de la Escosura-Muñiz et al. 2009; Henry et al. 2009). Importantly, an implantable biosensor has already been developed for cancer diagnosis in tissues (Daniel et al. 2009).

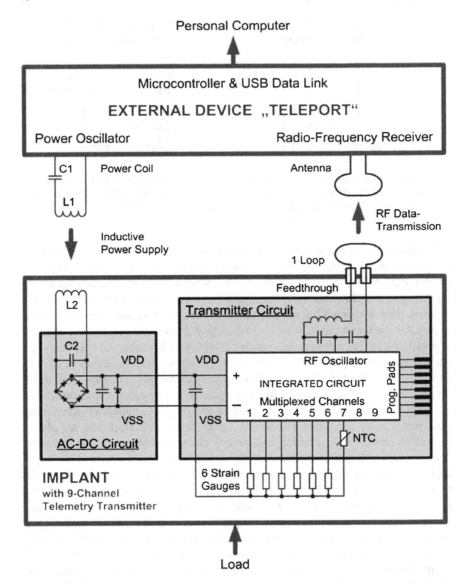

Fig. 3 Block diagram of the implantable telemetry system for measuring the implant load (adapted from Graichen et al. (2007))

However, to be a successful biosensor for surgical diagnosis, the main challenge is the target recognition material of the sensor. First of all, diagnosis agents (for example, DNA) must be immobilized onto biosensors with proper functionality and stability. Depending on the application and sensor material, a variety of approaches have been developed to synthesize and functionalize DNA onto biosensors.

This topic is stressed in Sect. 2 of this chapter. Moreover, for implantable biosensors in surgeries, a continuous concern is the biocompatibility of the sensor materials (Vaddiraju et al. 2010). Bio-incompatible sensors may result in the activation of the immune system, chronic inflammation, and formation of granulation tissue. Such reactions cause damage to the host tissue, as well as greatly decrease the lifetime and accuracy of the sensors. To overcome this gap, researchers intend to combine nanotechnology with bioengineering knowledge to produce novel DNA-derived materials with excellent biocompatibility and multifunctionality. This topic is discussed in Sect. 3. Therefore, the following chapter is filled with two types of materials for biosensor applications: natural DNA and DNA-derived materials, pointing out a way to combat current challenges and illustrating future prospects of DNA-based nanotechnology biosensors for surgical applications.

2 Natural DNA for Biosensors

DNA is known as important genetic material for all organisms since it is able to hybridize with its complimentary sequences in a highly selective way; this is also the advantage of DNA-based biosensors. Due to a variety of combinations of four different nucleotides, there are numerous DNA sequences. To functionalize DNA for biosensor applications, two steps are essential: synthesis and immobilization of DNA. Later in this section, both steps are detailed.

2.1 Synthesis of DNA

For controlling the structure (sequence) of defined nucleic acids used in biosensors, nucleic acids could be obtained by bond cleavage of longer segments, but they are now more commonly synthesized by chemically polymerizing individual nucleotide precursors. Chemical oligonucleotide synthesis is carried out from the 3' to 5' direction, which is opposite to enzyme synthesis. Synthesis starts from a solid support which is then processed by a step-by-step addition of nucleotide residues to the 5'-terminus of the elongating oligonucleotide chain until a desired sequence has been built. Here the general DNA/RNA synthesizer is taken as an example. Solid supports are nucleoside phosphoramidate (starting nucleotide residue) connected to controlled-pore glass via a carbon linker; they end up with a N,N-dimethyltryptamine (DMT) group which is essential to chain elongation. The following addition of each nucleotide residue is referred to as a synthetic cycle (Fig. 4), which consists of four basic reactions:

Step one, deblocking: the acid labile protecting group DMT is removed with an acid, such as trichloroacetic acid or dichloroacetic acid, in an inert solvent (dichloromethane or toluene) and washed out, resulting in a 5' free hydroxyl group ready for coupling.

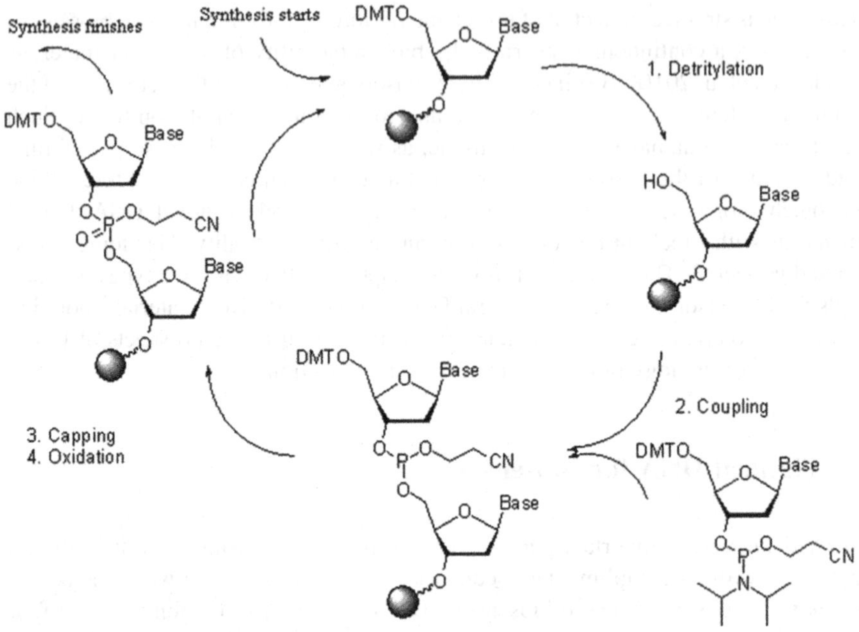

Fig. 4 A general DNA synthetic cycle

Step two, coupling: a nucleoside phosphoramidite is mixed with an acidic azole catalyst, 1H-tetrazole, 4,5-dicyanoimidazole, 2-ethylthiotetrazole, or other activators, brought in contact with the oligonucleotide precursor whose 5'-hydroxyl group is unprotected. The coupling reaction allows connecting of incoming nucleoside phosphoramidate to oligonucleotide precursors which are linked to a solid support, resulting in one nucleotide elongated phosphite triester. This coupling requires highly anhydrous conditions and often happens in anhydrous acetonitrile, which is an inert polar solvent. The final washing removes unconnected reagents and by-products.

Step three, capping: after the coupling reaction, there is still a small portion of solid support-bound 5' hydroxyl groups present. To prevent elongation of undesired sequences, those free 5' hydroxyl groups need to be blocked permanently. This reaction is done by treating the solid support with acetic anhydride/pyridine in tetrahydrofuran (THF) to form acetylation products of hydroxyl groups.

Step four, oxidation: phosphite triester generated in the coupling step is not naturally formed with confined stability under the condition following oligonucleotide elongation. Oxidation is achieved by treating the solid support-bound oligonucleotide with iodine water and pyridine in THF solution. The result is to transform phosphite trimesters into phosphate trimesters whose oxidation state is the same as native DNA.

Eventually, after the synthetic steps and purification steps, the designed DNA sequence is readily obtained.

2.2   Immobilization of DNA

Thus, immobilizing DNA to a variety of materials is a key parameter to produce DNA biosensors. Materials supporting DNA sequences in biosensors are often called transducers. Currently, there are three general types of immobilization that are widely used: adsorption, avidin–biotin complexation, and covalent attachment (Pividori et al. 2000).

2.2.1   DNA Immobilization by Adsorption

Adsorption is the most convenient method among those three mentioned above to immobilize nucleic acids on transducer surfaces. It does not require reagents or special nucleic acid sequences, but the affinity between the DNA and transducer surface is necessary. Suitable materials for this type of immobilization include polystyrene, metal oxides (such as aluminum oxide, palladium oxide), nylon membranes, and carbon/carbon paste transducers (Wang et al. 1996a, b, c).

Electrochemical adsorption on carbon paste surfaces is one of the most common DNA immobilization methods. There are plenty of reports to modify either single-strand DNA (ssDNA) or double-strand DNA (dsDNA) via a potential applied to carbon paste electrodes or to screen-printed carbon strip electrodes (Wang et al. 1996b, c; Wang et al. 1997a, b). For ssDNA, a voltage (vs. Ag/AgCl reference electrode) on a carbon transducer is applied for a couple minutes as a pretreatment. This pretreatment increases the roughness as well as hydrophilicity of the carbon surface (Wang et al. 1996a; Wang et al. 1997b). Afterwards, the electrochemical adsorption of ssDNA is implemented by stirring a solution on carbon surfaces at a potential for a preset time that depends on both the DNA sequence and concentration. The positive charge on carbon surfaces will enhance the stability of the immobilized DNA probe via electrostatic attraction. At the same time, such electrostatic attraction also increases the orientation of DNA probe alignment, so that a negative-charged hydrophilic sugar–phosphate backbone of the DNA probe can orient to carbon surfaces and the base part orients towards the solution readily hybridized with the target (Ronkainen et al. 2010; Palecek et al. 1998; Marrazza et al. 1999), as shown in Fig. 5. Hybridization will result in selective recognition and generation of biological signals for further diagnosis.

For dsDNA, no hybridization happens during the adsorption and detection process, so dsDNA electrochemical adsorption probes are suitable for non-DNA analytes, which can bind to dsDNA; those include both electroactive and nonelectroactive analytes. Double-strand DNA probes are easier to mount on electrodes (less orientation requirements), and due to a lack of the need for hybridization, they usually have longer lifetimes than ssDNA probes (Wang et al. 1997b).

Besides electrochemical adsorption, physical adsorption is also used in DNA sensors. Briefly, the DNA biosensor is prepared by dipping a glassy carbon electrode into a dsDNA solution leaving the electrode to dry (Oliveira Brett et al. 1997, 1998).

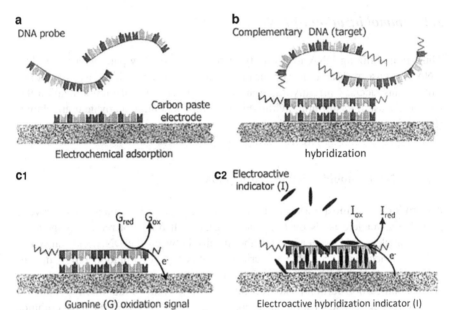

Fig. 5 DNA immobilization by electrochemical adsorption. (**a**) DNA probe adsorption to an electrochemical transducer applying a positive potential. (**b**) Hybridization between the probe and the target, holding the same positive potential. (**c**) Transduction. (**c1**) Transduction using the guanine oxidation signal. (**c2**) Indicator preconcentration in the dsDNA at a fixed positive potential and transduction based on the electroactive hybridization indicator (adapted from Pividori et al. (2000))

This sensor can be used in preconcentrating nitroimidazole or mitoxantrone on the surface and to study the interaction mechanism of these drug reactions with dsDNA by differential pulses, such as square wave voltammetry. Another method of physical dsDNA adsorption uses a glassy carbon disk. DNA-modified electrodes are prepared by the evaporation of a small volume of DNA solution on the electrode surface. The detection of the duplex was made with electroactive indicators ($Cu(phen)_2^{2+}$ and $Cu\text{-}(TAAB)^{2+}$) by cyclic voltammetric analysis of a preconcentrated electroactive indicator (Labuda et al. 1999).

Although adsorption is a simple and rapid method to immobilize DNA, the major disadvantage of this method is the labile connection between materials and DNA. As a result, DNA may dissociate from the transducer surface during hybridization conditions, causing low efficiency of hybridization. Thus, methods that rely on bond connections between DNA and transducer surface have been developed.

2.2.2 DNA Immobilization by Protein Complexation

Avidin or streptavidin–biotin complexation has been widely applied in the DNA biosensor field due to their high stability and high selectivity (Fig. 6). Avidin and

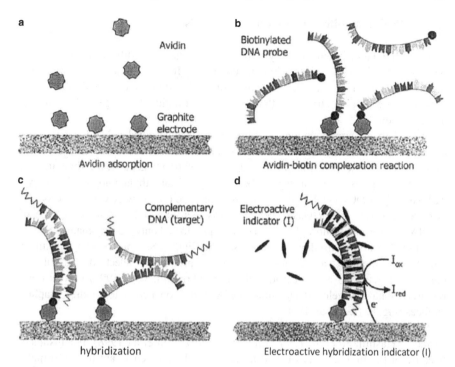

Fig. 6 DNA immobilization involving avidin–biotin complexation. (**a**) Avidin adsorption onto the graphite electrode. (**b**) Complexation between avidin and biotinylated DNA probe. (**c**) Hybridization of avidin–biotin probe with DNA target. (**d**) Signal transduction using an electroactive hybridization indicator preconcentrated into the dsDNA (adapted from Pividori et al. (2000))

streptavidin are tetramer proteins (70 kDa) incorporating four identical binding sites. Biotin is a small molecule that specifically binds with avidin or streptavidin ($Ka = 1 \times 10^{15}$ L/mol). Thus, this binding effect has a similar energy gain to a covalent bond, so it is quite stable and can only be removed under the most extreme conditions. Such robust bonds make the complex unaffected even in extreme values of pH, relative high temperature, organic solvents, and denaturing agents.

The tetravalent binding of streptavidin to biotin allows a general procedure for constructing DNA-coated electrodes, where the surface-bound avidin is coupled to a biotinylated oligonucleotide. This is followed by an avidin treatment to block any adsorption of DNA on a material surface (or proteins that provide a biotin head outside). The inherent aqueous stability of the avidin–biotin complex renders the system easy to handle. Because of the presence of a large protein layer, there might be some nonspecific interactions to block DNA–DNA binding sites (Pividori et al. 2000; Marrazza et al. 1999). Although usually such blocking would not significantly impact detection, due to the extremely high selectivity of hybridization, a more desired and direct approach has been investigated for DNA immobilization: covalent bond connections.

2.2.3 DNA Immobilization by Covalent Bonds

Covalent bond attachment is a very stable immobilization modification on materials with usually simple structures and mechanisms. It can provide stable linkages which allows for increasing hybridization times. In addition, chemical linkers will provide more flexible structures that allow conformational change without losing nucleic acids.

In 1993, the first modification was developed to install a covalent bond onto a glassy carbon surface (Millan et al. 1992; Millan and Mikkelsen 1993). In this case, the electrode was electrochemically oxidized in an acidic solution to generate a carboxylic acid on carbon surfaces. After rinsing, carbodiimide 1-(3-dimethylaminopropyl)-3-ethyl-carbodiimide (EDC) and N-hydro-xysulfosuccinimide (NHS) were evaporated on the electrode surface. The activated electrode was then exposed to a small volume of ssDNA and was again allowed to dry. These steps allowed only a deoxyguanosine residue of ssDNA immobilized on the carbon surface (Fig. 7). Several years later, a similar modification method was developed to modify carbon paste electrodes in bulk by either octadecylamine or stearic acid (Mikkelsen 1996; Millan et al. 1994). Furthermore, researchers have developed an amino–DNA bond connection to modify graphite surfaces (Fig. 8) (Liu et al. 1996).

Besides carbon-based materials, DNA can be also immobilized onto inorganic surfaces (Hashimoto et al. 1994; Yang et al. 1997; Sun et al. 1998). For instance, because of the high affinity of gold to thioalcohol, researchers are trying to make thiol-nucleic acids self-assemble with gold surfaces. Self-assembled aminoethanethiol -modified gold nanoparticles have been successfully connected to ssDNA.

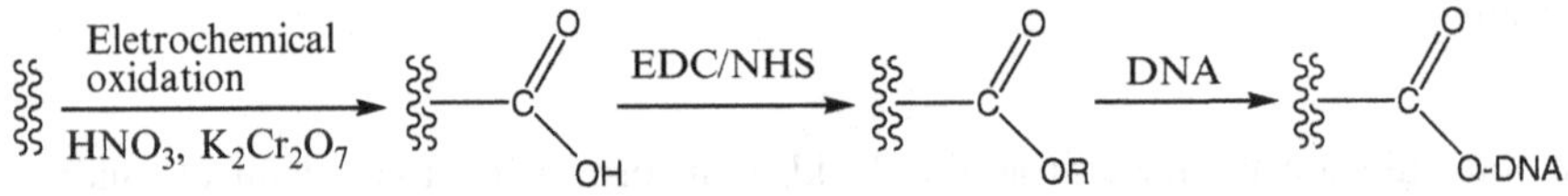

Fig. 7 Immobilization of ssDNA on glassy carbon electrodes

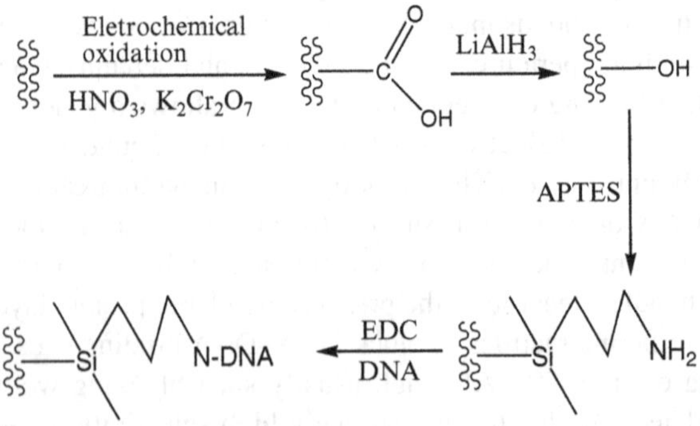

Fig. 8 Immobilization of ssDNA on graphite surfaces by using an amino–DNA linkage (*APTES* 3-aminopropyltriethoxysilane)

Moreover, ssDNA is added after generating a thiol-derivative monolayer to achieve a more precise way to control the coverage of gold surfaces. Such methods greatly reduce nonspecific DNA adsorption (Herne and Tarlov 1997; Steel et al. 1998). For polymer transducers, this methodology has been used to prepare immobilized DNA probes electrochemically using direct copolymerization of pyrrole and oligonucleotides bearing a pyrrole group (Livache et al. 1994). The process includes a direct electrochemical copolymerization of pyrrole and ssDNA with a pyrrole moiety introduced by phosphoramidate chemistry at their 5' end. Finally, a pyrrole group is covalently linked to a synthetic ssDNA.

In summary, adsorption is a fast, simple, and low-cost method to immobilize DNA onto biosensors. In contrast, covalent attachment usually needs more steps to build chemical bonds, but it is a highly tailorable method, which can connect with different materials. Therefore, depending on the specific application, a variety of methods are available for DNA immobilization.

3   DNA-Derived Materials for Biosensors

To obtain improved biocompatibility and multifunctionality for surgical diagnosis, researchers are engineering molecules to build novel DNA-analogous materials based on DNA structures. For example, from advances in nanotechnology, a promising biomaterial has emerged recently, which is a bioinspired, self-assembled nanomaterial. Such materials have excellent biological properties and are easy to engineer for specific needs. As one of the many examples, rosette nanotubes (RNTs, Fig. 9) are novel biomimetic self-assembled DNA-derived structures, whose basic building blocks are guanine (G) and cytosine (C) DNA base pairs which can solidify into a viscous gel (Chun et al. 2004, 2005; Fenniri et al. 2001). The G^C heteroaromatic bicyclic base possesses the Waston-Crick donor–donor–acceptor properties of guanine and the acceptor–acceptor–donor of cytosine. G^C undergoes a hierarchical self-assembly process under physiological conditions to form a six-membered supermacrocycle by the formation of 18 hydrogen bonds.

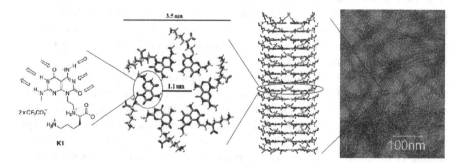

Fig. 9 RNTs (with a lysine side chain) undergo spontaneous self-assembly under physiological conditions (adapted and redrawn from Chun et al. (2004, 2005))

Because of electrostatic forces, base stacking interactions, and hydrophobic effects, the rosettes form a stable stack with an inner channel 11 Å in diameter. An amino acid side chain (lysine) was chosen to impose chirality and surface chemistry on the RNTs. Importantly, the lysine side chain (with an amine group and a carboxyl group) provides the possibility to functionalize a variety of peptides or proteins onto RNTs for enzymatic recognition. In addition, some small bioactive molecules could be trapped in the inner channel of RNTs by hydrogen bonds. This is another choice for the detection of small molecules using RNTs. In addition, RNTs have excellent cytocompatibility properties, suitable to be implanted into tissues for numerous sensor applications.

Therefore, RNTs not only have self-assembly and self-recognition properties (like DNA base pairs) but also have multiple functions, such as increasing viscosity when heated to low temperatures (40–60°C) and being able to connect with a high density of peptides or enzymes (Zhang et al. 2009). Moreover, RNTs are biocompatible with tissues because they can mimic the nanostructure of collagen and connective tissues to create a surface environment which improves protein (like collagen and fibronectin) adsorption as well as enhances cell adhesion and subsequent functions (Chun et al. 2004; Fenniri et al. 2001; Fine et al. 2009). RNTs are predicted to serve as a novel, multifunctional implant material for biosensors. In addition, in vivo toxicity studies have shown that RNTs lead to lower inflammatory responses than carbon nanotubes at the same dose (Journeay et al. 2008).

A previous study has further outlined a method to engineer RNTs with other materials (Chen et al. 2010a). Alginate gels, mixed with polyvinyl alcohol and polyethylene glycol, have been electrospun with RNTs. Transmission electron microscopy (TEM) imaging of this material clearly shows striations corresponding to RNTs on the electrospun fibers (arrows, Fig. 10a). In contrast, hydrogel composites without RNTs (Fig. 10b) were featureless. These observations suggested that the RNTs may behave as a novel nanoscaffolding on the surface of the hydrogel fibers with the ability of exposing bioactive functional groups (e.g., amino groups) to the surrounding tissues. Moreover, after electrospinning, the lysine side chains of the RNTs were labeled with a fluorescent probe to allow for the visualization of the RNT distribution in the hydrogel composite. Fluorescence microscopy imaging (Fig. 10c) showed that RNTs were present on the surface of the electrospun fibers, but not deep inside the fibers. In contrast, electrospinning fibers without RNTs (Fig. 10d) did not show any significant fluorescence. Importantly, these results confirmed the accessibility of the amino groups from the RNTs on the surface of the hydrogel fibers important for biosensor applications.

Furthermore, studies have highlighted the potential of RNTs to serve as a biomimetic nanoprobe for sensor applications. Figure 11a shows that HA crystals rapidly formed on RNTs when mixed with $CaCl_2$ and Na_2HPO_4 solutions for 2 min and were regularly aligned on RNTs in a pattern similar to the HA/collagen pattern in bone. Figure 11b shows randomly agglomerated HA on grids without RNTs (Zhang et al. 2008). Not only for ceramics, but metals have also been found to nucleate RNTs. Notably, the Au nanoparticles formed were nearly monodispersed clusters of gold particles (1.4–1.5 nm) nestled in pockets on the RNT surface

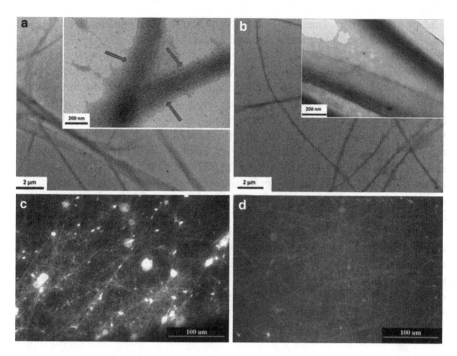

Fig. 10 Electrospun hydrogel fibers viewed in TEM (**a, b**) or under a fluorescence microscope (**c, d**). (**a, c**) with RNTs and (**b, d**) without RNTs. *Arrows* point to the RNTs on the hydrogel fiber surfaces (adapted and redrawn from Chen et al. (2010a))

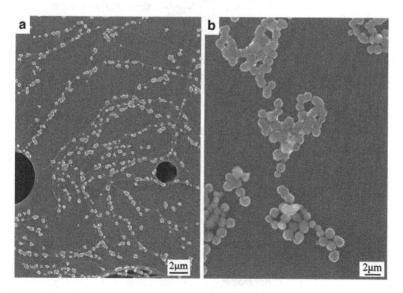

Fig. 11 Scanning electron microscope (SEM) images of (**a**) biomimetic HA with a 0.1 mg/mL RNT coating on a porous carbon TEM grid; and (**b**) HA coating on a TEM grid (adapted from Zhang et al. (2008))

(Chhabra et al. 2010). Thus, RNTs are very promising DNA-derived materials for biosensors and should be further explored in this context.

Besides RNTs, another DNA-derived molecule is a triplex-forming oligonucleotide (TFO). Normally, DNA is formed by two nucleotide strings. However, in another process, a triplex nucleic acid can be formed through the sequence-specific interaction between a single-stranded homopyrimidine and homopurine TFO (Chen et al. 2010b; Duca et al. 2008). In this strategy, a TFO binds to the major groove of the target duplex in a sequence-specific manner through hydrogen bonding between its bases and exposed groups on the duplex base pairs, generating base triplets (Fig. 12). TFO has been the focus of considerable interest because of possible applications in developing new molecular biology tools as well as diagnostic agents (Duca et al. 2008; Igoucheva et al. 2004; Buchini and Leumann 2003).

An important application of TFO as a biosensor material is the recognition of nucleic acids. A big advantage is the fact that pyrimidine triplex formation is pH-dependent. Thus, the recognition of TFO is quite sensitive on the acidity of the environment and such a process is tunable. Triplexes can be used to select a sequence of DNA containing oligopurine/oligopyrimidine sequences in a mixture of duplex DNA (Duca et al. 2008; Ji and Smith 1993). Previous studies have realized this idea

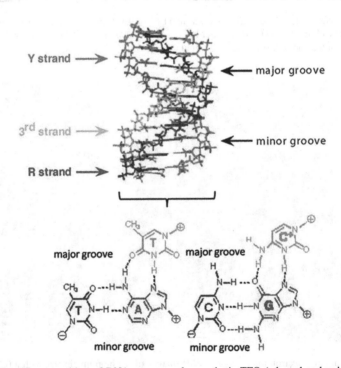

Fig. 12 Specific recognition of DNA sequences by synthetic TFO (adapted and redrawn from Duca et al. (2008))

by attaching a TFO to magnetic probes. In the presence of a mixture of duplexes (such as plasmids), a sequence able to form a triple helix can be selectively recognized inside a bacterial lysate (Ito et al. 1992; Schluep and Cooney 1998).

In another study, TFOs were combined with gold nanoparticles to detect DNA by a color change. Briefly, the pro-synthesized TFOs were adsorbed onto gold nanoparticles to prevent the individual red gold nanoparticles from forming blue aggregations under high-ionic strength conditions due to the low stability of the triplex structure. However, introducing a triplex binder/inducer stabilizes triplex formation through Hoogsteen-type hydrogen bonds. Thus, the triplex would not be able to bind and stabilize individual red gold nanoparticles, resulting in purple–blue gold nanoparticle aggregations (Fig. 13 (Chen et al. 2010b)).

In summary, DNA-derived materials have emerged for biosensor applications. They usually present unique properties, such as excellent biocompatibility (like RNTs) and high sensitivity for biomolecule recognition (like TFO). Therefore, based on nanotechnology and bioengineering, DNA-derived materials are outstanding candidates for a new generation of biosensors more effective for diagnosis.

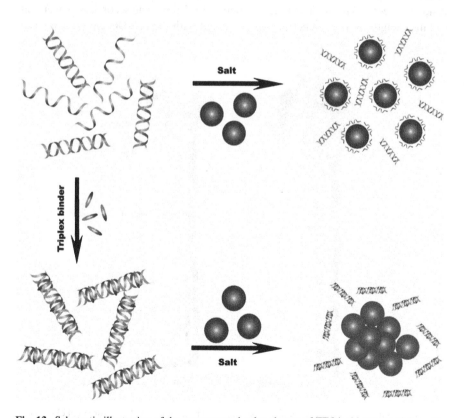

Fig. 13 Schematic illustration of the structure and color change of TFO/gold nanoparticles with the detection of target DNA (adapted from Chen et al. (2010b))

4   Future Prospects and Conclusions

Previously, biosensors were successfully used only in laboratory analysis, but today, the need of biosensors for in situ diagnosis is growing rapidly. Especially, because of the high selectivity, sensitivity, and generally good biocompatibility, DNA-based biosensors have a bright future for surgical applications. Along this route, multifunctional materials and composite structures are the future for DNA-based biosensors. An ideal biosensor should not only complete its intended function (recognition and detection), but it also needs to possess other benefits (such as drug release, mechanical support, gene therapy, biocompatibility, etc.) (Ménard-Moyon et al. 2010; Ainslie and Desai 2008; Wadhwa et al. 2006).

Take a currently developed bone sensor for example. An implantable 9-channel implant/biosensor system was studied in vivo for load measurements for orthopedic implants (Graichen et al. 2007). Briefly in the experiment, a Ti alloy (Ti4Al6V) was used for the neck and stem, cobalt-chromium-molybdenum (CoCrMo) for the tightly fitted head, and polyether ether ketone (PEEK) for the cap at the lower end of the stem (Fig. 14). Six strain gauges and the transmitter circuit were arranged inside the neck cavity with an inner diameter of 9.5 mm. The secondary power coil (30 mm length, 6 mm diameter) was combined with the AC-DC circuit and fixed

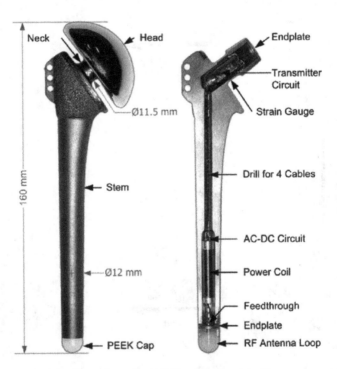

Fig. 14 Instrumented shoulder endoprosthesis (*left*) and a cut model with measurement electronics (*right*) (adapted from Graichen et al. (2007))

in a second cavity at the lower end of the stem. The rectified power supply voltages were connected to the transmitter circuit by two cables through a drill. Another pair of cables connected the two-lead feed through at the center of the lower end plate to the transmitter. The antenna loop for signal transmission consisted of a niobium wire. The wire was electron-beam welded and protected against mechanical damage by a PEEK cap, and the upper and lower endplates of the implant were also welded. This is quite a novel design to incorporate a biosensor into a normal orthopedic implant. A significant advantage of this composite system is that a sensor can be located inside the patient's body. Thus, "real time" measuring of implant performance could be accomplished by the sensor. Compared with traditional methods, like radiographs (X-rays), such biosensor system results may detect implant failures earlier and more quickly. It would also lead to less radiation to patients and a more accurate diagnosis for surgeons.

In the future, more functions can be added into such systems. For example, RNTs can be incorporated onto the implants to improve biocompatibility, while TFO can be immobilized onto the implant surface to recognize certain biomarkers and analyze cell differentiation, functions, and apoptosis to predict the outcome of the implant before clinical symptoms surface. Furthermore, controlled drug release mechanisms could also be included. A conductive polymer, like polypyrrole, can be coated on the implants and embedded with drugs. With electrical stimulation, molecular chains of polypyrrole break and can release the embedded drugs (Ainslie and Desai 2008; Wadhwa et al. 2006). Notably, by controlling the electrical current in the implant, the drug release rate is adjustable. In this manner, the diagnosis and treatment of the implant can be achieved simultaneously through such novel DNA-based nanotechnology biosensors.

In conclusion, DNA or DNA-derived materials are very promising for biosensor applications to improve their biocompatibility, selectivity, and sensitivity of the sensors, and combining with other functions (such as drug release and gene therapy) to provide earlier, real-time diagnosis and treatment.

References

Ainslie KM, Desai TA. Microfabricated implants for applications in therapeutic delivery, tissue engineering, and biosensing. Lab Chip. 2008;8(11):1864–78.

Bi H, Meng S, Li Y, Guo K, Chen Y, Kong J, Yang P, Zhong W, Liu B. Deposition of PEG onto PMMA microchannel surface to minimize nonspecific adsorption. Lab Chip. 2006; 6(6):769–75.

Buchini S, Leumann CJ. Recent improvements in antigene technology. Curr Opin Chem Biol. 2003;7:717–26.

Chen Y, Bilgen B, Pareta RA, Myles AJ, Fenniri H, Ciombor DM, Aaron RK, Webster TJ. Self-assembled rosette nanotube/hydrogel composites for cartilage tissue engineering. Tissue Eng Part C Methods. 2010a. Epub ahead of print.

Chen C, Song G, Yang X, Ren J, Qu X. A gold nanoparticle-based strategy for label-free and colorimetric screening of DNA triplex binders. Biochimie. 2010b;92(10):1416–21.

Chhabra R, Moralez JG, Raez J, Yamazaki T, Cho JY, Myles AJ, Kovalenko A, Fenniri H. One-pot nucleation, growth, morphogenesis, and passivation of 1.4 nm Au nanoparticles on self-assembled rosette nanotubes. J Am Chem Soc. 2010;132(1):32–3.

Chun AL, Moralez JG, Fenniri H, Webster TJ. Helical rosette nanotubes: a more effective. orthopaedic implant material. Nanotechnology. 2004;15:s234–9.

Chun AL, Moralez JG, Fenniri H, Webster TJ. Helical rosette nanotubes: a biomimetic coating for orthopedics? Biomaterials. 2005;26:7304–9.

Daniel KD, Kim GY, Vassiliou CC, Galindo M, Guimaraes AR, Weissleder R, Charest A, Langer R, Cima MJ. Implantable diagnostic device for cancer monitoring. Biosens Bioelectron. 2009;24(11):3252–7.

de la Escosura-Muñiz A, Sánchez-Espinel C, Díaz-Freitas B, González-Fernández A, Maltez-da Costa M, Merkoçi A. Rapid identification and quantification of tumor cells using an electrocatalytic method based on gold nanoparticles. Anal Chem. 2009;81(24):10268–74.

Drummond TG, Hill MG, Barton JK. Electrochemical DNA sensors. Nat Biotechnol. 2003;21(10):1192–9.

Duca M, Vekhoff P, Oussedik K, Halby L, Arimondo PB. The triple helix: 50 years later, the outcome. Nucl Acids Res. 2008;36:5123–38.

Durr HR, Maier M, Jansson V, Baur A, Refior HJ. Phenol as an adjuvant for local control in the treatment of giant cell tumour of the bone. Eur J Surg Oncol. 1999;25(6):610–18.

Fenniri H, Mathivanan P, Vidale KL, Sherman DM, Hallenga K, Wood KV, Stowell JG. Helical rosette nanotubes: design, self-assembly and characterization. J Am Chem Soc. 2001;123:3854–5.

Fine E, Zhang L, Fenniri H, Webster TJ. Enhanced endothelial cell functions on rosette nanotube-coated titanium vascular stents. Int J Nanomed. 2009;4:91–7.

Graichen F, Bergmann G. Four channel telemetry system for in vivo measurement of hip joint forces. J Biomed Eng. 1991;13(5):370–4.

Graichen F, Bergmann G, Rohlmann A. Hip endoprosthesis for in vivo measurement of joint force and temperature. J Biomech. 1999;32(10):1113–17.

Graichen F, Arnold R, Rohlmann A, Bergmann G. Implantable 9-channel telemetry system for in vivo load measurements with orthopedic implants. IEEE Trans Biomed Eng. 2007;54(2):253–61.

Guo Y, Ye JY, Divin C, Huang B, Thomas TP, Baker JR Jr, Norris TB. Real-time biomolecular binding detection using a sensitive photonic crystal biosensor. Anal Chem. 2010; 82(12):5211–8.

Hashimoto K, Ito K, Ishimori Y. Sequence-specific gene detection with a gold electrode modified with DNA probes and an electrochemically active dye. Anal Chem. 1994;66:3830–3.

Henry OY, Fragoso A, Beni V, Laboria N, Sánchez JL, Latta D, Von Germar F, Drese K, Katakis I, O'Sullivan CK. Design and testing of a packaged microfluidic cell for the multiplexed electrochemical detection of cancer markers. Electrophoresis. 2009;30(19): 3398–405.

Herne TM, Tarlov MJ. Characterization of DNA probes immobilized on gold surfaces. J Am Chem Soc. 1997;119:8916–20.

Hoch DB, Zieve D. Electromyography. Medical Encyclopedia. National Library of Medicine. 2008. Available at: http://www.nlm.nih.gov/medlineplus.

Huang H, Li J, Tan Y, Zhou J, Zhu JJ. Quantum dot-based DNA hybridization by electrochemiluminescence and anodic stripping voltammetry. Analyst. 2010;135(7):1773–8.

Igoucheva O, Alexeev V, Yoon K. Oligonucleotide-directed mutagenesis and targeted gene correction: a mechanistic point of view. Curr Mol Med. 2004;4:445–63.

Ito T, Smith CL, Cantor CR. Sequence-specific DNA purification by triplex affinity capture. Proc Natl Acad Sci U S A. 1992;89:495–8.

Ji H, Smith LM. Rapid purification of double-stranded DNA by triple-helix-mediated affinity capture. Anal Chem. 1993;65:1323–8.

Journeay WS, Suri SS, Moralez JG, Fenniri H, Singh B. Rosette nanotubes show low acute pulmonary toxicity in vivo. Int J Nanomed. 2008;3(3):373–83.

Labuda L, Bučková M, Vaníčková M, Mattusch L, Wennrich R. Voltammetric detection of the DNA interaction with copper complex compounds and damage to DNA. Electroanalysis. 1999;11:101–7.

Leegsma-Vogt G, Rhemrev-Boom MM, Tiessen RG, Venema K, Korf J. The potential of biosensor technology in clinical monitoring and experimental research. Biomed Mater Eng. 2004;14(4):455–64.

Lin TW, Kekuda D, Chu CW. Label-free detection of DNA using novel organic-based electrolyte-insulator-semiconductor. Biosens Bioelectron. 2010;25(12):2706–10.

Liu S, Ye L, He P, Fang Y. Voltammetric determination of sequence-specific DNA by electroactive intercalator on graphite electrode. Anal Chim Acta. 1996;335:239–43.

Livache T, Roget A, Dejean E, Barthet C, Bidan G, Te´oule R. Preparation of a DNA matrix via an electrochemically directed copolymerization of pyrrole and oligonucleotides bearing a pyrrole group. Nucl Acids Res. 1994;22:2915–21.

Marrazza G, Chianella L, Mascini M. Disposable DNA electrochemical sensor for hybridization detection. Biosens Bioelectron. 1999;14:43–51.

McLaughlin KJ, Strain-Damerell CM, Xie K, Brekasis D, Soares AS, Paget MS, Kielkopf CL. Structural basis for NADH/NAD+ redox sensing by a Rex family repressor. Mol Cell. 2010;38(4):563–75.

McNaught AD, Wilkinson A. International Union of Pure and Applied Chemistry, Second Edition. Compendium of Chemical Terminology. Blackwell Science, 1997.

Ménard-Moyon C, Kostarelos K, Prato M, Bianco A. Functionalized carbon nanotubes for probing and modulating molecular functions. Chem Biol. 2010;17(2):107–15.

Mikkelsen SK. Electrochemical biosensors for DNA sequence detection. Electroanalysis. 1996;8:15–19.

Millan KM, Mikkelsen SK. Sequence-selective biosensor for DNA based on electroactive hybridization indicators. Anal Chem. 1993;65:2317–23.

Millan KM, Spurmanis AL, Mikkelsen SK. Covalent immobilization of DNA onto glassy carbon electrodes. Electroanalysis. 1992;4:929–32.

Millan KM, Saraullo A, Mikkelsen SK. Voltammetric DNA biosensor for cystic fibrosis based on a modified carbon paste electrode. Anal Chem. 1994;66:2943–8.

Oliveira Brett AM, Serrano SHP, Gutz I, La-Scalea MA, Cruz ML. Voltammetric behavior of nitroimidazoles at a DNA-biosensor. Electroanalysis. 1997;9:1132–7.

Oliveira Brett AM, Macedo TRA, Raimundo D, Marques MH, Serrano SHP. Voltammetric behaviour of mitoxantrone at a DNA-biosensor. Biosens Bioelectron. 1998;13:861–7.

Palecek R, Fojta M, Tomschik M, Wang L. Electrochemical biosensors for DNA hybridization and DNA damage. Biosens Bioelectron. 1998;13:621–8.

Pividori MI, Merkoçi A, Alegret S. Electrochemical genosensor design: immobilisation of oligonucleotides onto transducer surfaces and detection methods. Biosens Bioelectron. 2000;15(5–6):291–303.

Ronkainen NJ, Halsall HB, Heineman WR. Electrochemical biosensors. Chem Soc Rev. 2010;39(5):1747–63.

Schluep T, Cooney CL. Purification of plasmids by triplex affinity interaction. Nucl Acids Res. 1998;26:4524–8.

Steel AR, Herne TM, Tarlov MJ. Electrochemical quantification of DNA immobilized on gold. Anal Chem. 1998;70:4670–7.

Storhoff JJ, Elghanian R, Mucic RC, Mirkin CA, Letsinger RL. One-pot colorimetric differentiation of polynucleotides with single base imperfections using gold nanoparticle probes. J Am Chem Soc. 1998;120:1959–64.

Sun X, He P, Liu S, Ye L, Fang Y. Immobilization of single-stranded deoxyribonucleic acid on gold electrode with self-assembled aminoethanethiol monolayer for DNA electrochemical sensor applications. Talanta. 1998;47:487–95.

Vaddiraju S, Tomazos I, Burgess DJ, Jain FC, Papadimitrakopoulos F. Emerging synergy between nanotechnology and implantable biosensors: a review. Biosens Bioelectron. 2010;25(7):1553–65.

Wadhwa R, Lagenaur CF, Cui XT. Electrochemically controlled release of dexamethasone from conducting polymer polypyrrole coated electrode. J Control Release 2006;110(3):531–41.

Wang J, Cai X, Jonsson C, Balakrishnan M. Adsorptive stripping potentiometry of DNA at electrochemically pretreated carbon paste electrodes. Electroanalysis. 1996;8:20–4.

Wang J, Cai X, Rivas G, Shiraishi H. Stripping potentiometric transduction of DNA hybridization processes. Anal Chim Acta. 1996;326:141–7.

Wang J, Cai X, Rivas G, Shiraishi H, Farias PAM, Dontha N. DNA electrochemical biosensor for the detection of short DNA sequences related to the human immunodeficiency virus. Anal Chem. 1996;68:2629–34.

Wang J, Cai X, Fernandes JR, Grant DH, Ozsoz M. Electrochemical measurements of oligonucleotides in the presence of chromosomal DNA using membrane-covered carbon electrodes. Anal Chem. 1997;69:4056–9.

Wang J, Cai X, Rivas G, Shiraishi H, Dontha N. Nucleic-acid immobilization, recognition and detection at chronopotentiometric DNA chips. Bioelectron. 1997;12:587–99.

Wei F, Lillehoj PB, Ho CM. DNA diagnostics: nanotechnology-enhanced electrochemical detection of nucleic acids. Pediatr Res. 2010;67(5):458–68.

Wilkins E, Atanasov P. Glucose monitoring: state of the art and future possibilities. Med Eng Phys. 1996;18(4):273–88.

Xia F, Zuo X, Yang R, Xiao Y, Kang D, Vallée-Bélisle A, Gong X, Yuen JD, Hsu BB, Heeger AJ, Plaxco KW. Colorimetric detection of DNA, small molecules, proteins, and ions using unmodified gold nanoparticles and conjugated polyelectrolytes. Proc Natl Acad Sci USA. 2010;107(24):10837–41.

Yang M, McGovern ME, Thompson M. Genosensor technology and the detection of interfacial nucleic acid chemistry. Anal Chim Acta. 1997;346:259–75.

Zhang L, Chen Y, Rodriguez J, Fenniri H, Webster TJ. Biomimetic helical rosette nanotubes and nanocrystalline hydroxyapatite coatings on titanium for improving orthopedic implants. Int J Nanomed. 2008;3(3):323–33.

Zhang L, Rakotondradany F, Myles AJ, Fenniri H, Webster T. Arginine-glycine-aspartic acid modified rosette nanotube-hydrogel composites for bone tissue engineering. Biomaterials. 2009;30(7):1309–20.

Chapter 5
Electrically Active Neural Biomaterials

Justin T. Seil and Thomas J. Webster

Abstract Numerous biomaterials have provided promising results toward improving the function of nervous system tissue. However, significant hurdles, such as delayed or incomplete tissue regeneration, remain toward full functional recovery of peripheral and central nervous system (CNS) tissue. Because of this continual need for better nervous system biomaterials, more recent approaches to design the next generation of tissue engineering scaffolds for the nervous system have incorporated nanotechnology, or more specifically, nanoscale surface feature dimensions which mimic that of the natural neural tissue. Compared to conventional materials with micron scale surface dimensions, nanomaterials have exhibited an ability to enhance desirable neural cell activity while minimizing unwanted cell activity, such as reactive astrocyte activity in the CNS. The complexity of neural tissue injury and the presence of inhibitory cues as well as the absence of stimulatory cues may require multifaceted treatment approaches with customized biomaterials that nanotechnology can provide. A combination of stimulatory cues may be used to incorporate nanoscale topographical and chemical or electrical cues in the same scaffold to provide an environment for tissue regeneration that is superior to inert scaffolds. Ongoing research in the field of electrically active nanomaterials includes the fabrication of composite materials with nanoscale, piezoelectric zinc oxide particles embedded into a polymer matrix. Zinc oxide, when mechanically deformed through ultrasound, for example, can theoretically provide an electrical stimulus, a known stimulatory cue for neural tissue regeneration. The combination of nanoscale surface dimensions and electrical activity may provide an environment for enhanced neural tissue regeneration; such multifaceted nanotechnology approaches deserve further attention in the neural tissue regeneration field.

Keywords Nanomaterials • Nanoparticles • Neural tissue regeneration • Zinc oxide • Piezoelectric

T.J. Webster (✉)
School of Engineering, Brown University, 182 Hope Street, Providence, RI 02917, USA
e-mail: thomas_webster@brown.edu

T.J. Webster (ed.), *Nanotechnology Enabled In situ Sensors for Monitoring Health*,
DOI 10.1007/978-1-4419-7291-0_5, © Springer Science+Business Media, LLC 2011

1   Introduction: Nervous System Injury

While injuries to the peripheral nervous system (PNS) are often capable of spontaneously healing after traumatic injury, damaged central nervous system (CNS) tissue does not regenerate in the same manner. In spite of decades of progress toward
restoring motor and sensory function to those with CNS injury, a complete solution
remains elusive to date. The current treatment of CNS injury, particularly injury to
the spinal cord, relies on minimizing secondary injury and implementing physical
therapy designed to help a patient function with limited mobility. Treatment to
restore healthy tissue and regain sensory and motor function is still not a reality.
Developments in CNS tissue regeneration aim to ultimately provide a method for
repairing functional tissue and restoring sensory and motor function.

1.1   Statistics

In the United States, a quarter of a million people live with a spinal cord injury.
Worldwide, approximately two million people live with a spinal cord injury (National
Spinal Cord Injury Statistical Center 2007). A majority of people with spinal cord
injury are below the age of 30 at the time of injury and, as a result, may require many
years of treatment or assisted living. Estimates place the number of new spinal cord
injuries each year at around 13,000. Depending on the severity and location of CNS
injury, motor and sensory function can be mildly affected in the lower extremities or
entirely eliminated below the neck. Many attempts have been made to develop
therapies for CNS regeneration, but they have been met with limited success. These
difficulties make a successful treatment very desirable due to the life-changing impact
that CNS injury has on the lives of millions.

1.2   Central Nervous System Injury

The lack of tissue regeneration in the CNS is due to a complex series of events
that follow injury. Though these events are necessary for the restoration of the
blood–brain barrier and minimization of secondary tissue damage, they also
result in the formation of an environment which is not conducive to tissue regeneration. Damage to the CNS initiates an astrocytic response resulting in a glial
scar. Damaged tissue releases molecular cues which promote astrocytic glial cell
activity. Transforming growth factor β (TGFβ) is among the molecular cues that
increase immediately after injury and contribute to the initiation of reactive
gliosis. Astrocytes migrate, proliferate, increase in size, and produce a glial scar
rich in extracellular matrix (ECM) proteins, myelin, astrocytes, and oligodendrocytes. While hypertrophic astrocytes comprise a significant portion of the glial
scar, production of glycosaminoglycans is critical to glial scar formation.
Reactive astrocytes produce a variety of proteoglycans including chondroitin

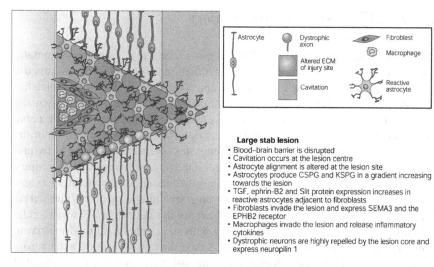

Fig. 1 Illustration of CNS glial scarring after injury. A series of events leads to the formation of an environment that is not conducive to tissue regeneration. Among these events, reactive astrocytes form an inhibitory glial scar. However, astrocytes have exhibited reduced activity on select nanomaterials. Figure adapted from (Silver and Miller 2004)

sulfate proteoglycans (CSPGs). CSPGs are known to aggressively inhibit axon outgrowth (Silver and Miller 2004). Glial scar tissue can prevent spontaneous neural tissue regeneration by serving as a physical and chemical barrier to tissue regeneration (Fig. 1). Axonal outgrowth from healthy tissue is impeded by this glial scar, preventing reinnervation of targets on the other side of the injury. Without a clear pathway for growing axons, neural tissue regeneration is impossible.

Beyond traumatic injury to the spine, damage as minimal as that which is caused by the insertion of a microelectrode into the brain can initiate an astrocytic response. Upon insertion of a microelectrode, neural cell tissue, as well as the blood–brain barrier, is damaged. To repair broken blood vessels, support damaged neural cells, and clean up cellular debris, microglia, astrocytes, and macrophages are recruited to the site of injury by molecular signaling. Though this cellular activity is necessary to repair the blood–brain barrier and minimize secondary damage, the result of overactive astrocytes is the formation of glial scar tissue similar to that which forms following traumatic injury. In this case, the formation of glial scar tissue compromises the effectiveness of the microelectrode. This chronic inflammatory response is a major complication for electrodes implanted for extended periods of time (Polikov et al. 2005). The utility of the microelectrode depends on a conductive interface to record or deliver electrical impulses. As the glial scar forms, the impedance at the electrode interface increases. Significantly reduced conductivity can render the electrode ineffective. Both spinal cord injury and microelectrode insertion result in compromised CNS tissue. Biomaterials, whether used for the fabrication of a microelectrode or formed into a scaffold for CNS tissue regeneration, aim to moderate astrocytic glial cell activity while promoting neural tissue growth. This is the promise for nanotechnology and nanomaterials.

1.3   Peripheral Nervous System Injury

Peripheral nerve damage is typically, but not exclusively, caused by traumatic injury. Nerves can also be damaged by other means, including orthopedic surgery complications. For example, in approximately 1–3% of all total hip replacements, the sciatic nerve can be injured by excessive tension when the extremity has been lengthened significantly (Wasielewski et al. 1992). If a nerve is transected by severe injury, the nerve function is lost. Thus, the portion of the nerve distal to the injury dies and degenerates. Nevertheless, the proximal segment may be able to regenerate and reestablish nerve function. The injury creates a gap which must be bridged by the growth cones of regenerating axons. First, Schwann cells and macrophages remove myelin and other cellular debris from the injury site. A fibrin cable forms across the injury gap. Schwann cells then infiltrate and orient themselves, along with an oriented laminin-1 matrix, to form the aligned Bands of Bungner which guide the newly forming axon across the site of injury. PNS tissue does not contain astrocytes and is not complicated by glial scarring, however, granulation tissue formation rather than nerve regeneration has been observed.

2   Nerve Repair Strategies

To repair severed nerves, the stump ends of the nerve fiber bundles can be surgically reconnected and sutured together. Transected tissue will generally not regenerate without surgical realignment (Wasielewski et al. 1992). A tissue transection, with additional tissue damage that creates a gap in the tissue, may not be a candidate for surgical reconnection if the reconnection will create tension in the nerve. Thus, large gaps must be repaired with a graft inserted between the proximal and distal nerve stumps as a guide for regenerating axons.

While the current gold standard of treatment is the use of an autologous sural nerve graft, there is a great deal of interest in the development of synthetic nerve guidance channels (NGC; Fig. 2) that perform equally well. Treatment of nerve fiber transections and small tissue gaps, has a high rate of success with both synthetic and autologous implants. However, as the tissue gap increases in length the probability of successful regeneration decreases dramatically (Yannas and Hill 2004). NGC technology is therefore of interest due to the potential inclusion of known stimuli which may allow regeneration of larger tissue gaps, and due to the complications associated with producing an autologous graft. The process of harvesting an autologous graft requires an additional surgical operation, may result in sensory complications at the donor site, and exposes the patient to increased surgical risks. The NGC is, in its most basic form, a hollow tube into which two severed ends of a nerve fiber bundle can be inserted and sutured into place. The implanted

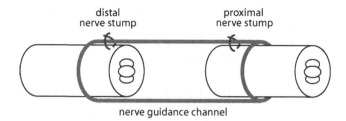

distal   proximal
nerve stump   nerve stump

nerve guidance channel

Fig. 2 Illustration of a NGC bridging two severed ends of a transected nerve fiber bundle. Though no commercially available NGCs are fabricated from electrically-active materials, future developments in conductive or piezoelectric materials for NGC applications may accelerate axon growth between nerve stumps

NGC serves to protect and support regenerating tissue. The NGC assists the realignment of nerve tissue and maintains high levels of locally released growth factors by encapsulating the nerve.

While degradable polymer scaffolds without biologically-inspired nanoscale features are, to date, the standard NGC material choice, studies have investigated the efficacy of NGC materials with a wide variety of properties. Commercially available NGCs (Table 1), such as the NeurgaGen collagen tube (Integra LifeSciences, Plainsboro, New Jersey) and the hydrogel-based SaluBridge (SaluMedica, Atlanta, Georgia) are inert polymer tubes that do not incorporate the several neural growth stimulatory cues under active investigation (Meek and Coert 2008). Standard NGCs do incorporate desirable flexibility, which is an important element in NGC design due to the need to reduce tension and mechanical stress during nervous system tissue regeneration (Wasielewski et al. 1992). Also, biodegradation is considered to be an essential quality for NGC materials. As tissue regenerates, the scaffold will ideally degrade as it is replaced by healthy tissue. Failure of the scaffold to degrade could potentially constrict growing tissue and prevent proper regeneration. Also, degradable NGCs eliminate the need for a second operation for implant removal.

2.1 Repair Strategies in the PNS

While damaged tissue in the PNS is able to spontaneously regenerate after injury, the regeneration rate is quite slow. Axonal growth rates have been estimated at about 0.5–1 mm/day (Archibald et al. 1995). Although those growth rates may translate into the bridging of a several millimeter nerve gap in a matter of days, reinnervation of target tissue can take months. Accelerating recovery is desirable due to the loss of function prior to appreciable tissue regeneration. The slow growth of axons delays the functional recovery of PNS tissue and allows for extended periods of muscle atrophy. A number of techniques used to accelerate tissue regeneration

Table 1 Commercially available, FDA-approved NGCs

Product name	NeuraGen	NeuroMatrix NeuroFlex	Neurotube	Neurolac	Salubridge
Material	Type I collagen	Type I collagen	Polyglycolic acid	Poly (85:15/L:D) lactide-[varepsilon]-caprolactone	Elastomer-hydrogel
Maximum length	3 cm	2.5 cm	4 cm	3 cm	6.35 cm
Diameter	1.5–7 mm	2–6 mm	2.3–8 mm	1.5–10 mm	2–10 mm
Degradation time	4–8 months	4–8 months	6 months	16 months	Does not degrade
FDA approval date	6/2001	9/2001	3/1999	10/2003	Late 2001

have been identified, such as electrical stimulation (Rajnicek et al. 1998; Borgens et al. 1999) and the release of a variety of growth factors such as brain-derived neurotrophic factor (BDNF) in the CNS (Bregman et al. 1997) and glial cell-line derived neurotrophic factor (GDNF) in the PNS (Henderson et al. 1994). Second, if the tissue damage is great enough, a critical defect incapable of spontaneously regenerating is produced. Both of these scenarios may benefit from the implantation of a NGC to permit or accelerate the regeneration process. It is important to note that mature neurons do not undergo mitosis. Thus, supporting the regrowth of axons from existing cells to distal targets is the goal of a NGC.

2.2 Repair Strategies in the CNS

A NGC in the CNS must reduce astrocyte activity, enhance neural cell activity and axonal outgrowth, contribute to maintaining high local concentrations of desirable growth factors, and support transplanted cells including Schwann cells, olfactory ensheathing cells, oligodendrocytes, or neural stem cells. The major point to consider when designing a NGC for the CNS is the difference in the presence and activity of glial cells. Astrocyte activity is not an issue in the PNS due to the absence of astrocytes from that tissue type. However, the presence of Schwann cells in the PNS, and the role that they play in providing an environment for successful regeneration, must be taken into consideration in developing methods for enhancing regeneration in the CNS.

3 Metrics of Regeneration

The success and effectiveness of the strategies investigated for neural tissue regeneration have been measured in a variety of ways in vitro and in vivo. In the PNS, an ideal material for implantation should optimize neural and Schwann cell function. In the CNS, the materials should optimize neural cell function but also minimize the excessive glial scar tissue produced by reactive astrocytes. Metrics for evaluating cell activity, associated proteins, tissue structure, and sensory and motor function have been used to analyze the success of the variety of biomaterials and treatment techniques.

The response of a variety of relevant cell types to a range of biomaterials has been evaluated. Animal models have provided more physiological data to identify some promising approaches to neural tissue engineering.

3.1 Neural Cells

In vitro, the response of neural and associated cells has been evaluated on a variety of materials. Multiple neural and neural-like cell types have been used to evaluate cell response to the scaffold on which they are grown. One such cell type, PC12, originates from the adrenal glands. Although this cell type is commonly chosen for neural tissue engineering experiments due to its relative ease of use and thorough characterization, it is not as physiologically relevant as other neural cell options and it does not produce true axons and dendrites. Nevertheless, it has been used in multiple studies to indicate desirable scaffold properties. Cell counts and increased number and length of neurites, extensions from the cell, are accepted as indicators of positive PC12 cell responses.

Dorsal root ganglion (DRG) cells are more physiologically relevant in comparison to PC12 cells. However, the collection and culture process is more demanding. Primary DRGs are harvested from the spinal cord of sacrificed Sprague-Dawley rats at around embryonic-day 15 (Corey et al. 2007). Cells may either be placed on uncoated samples or samples onto which laminin or other proteins have been adsorbed. As with PC12 cells, cell number, but particularly neurite number and length, are evaluated. Studies which have incorporated a directional stimulus may also evaluate the direction of neurite growth on the substrate. Antibodies for axon neurofilaments (Corey et al. 2007) or proteins associated with growth cone development can also be used as an indicator of cell activity.

3.2   Glial Cells

Reactive astrocyte activity on a material is an indicator of how well the material will minimize undesirable glial scarring after implantation. This is relevant for CNS applications of neural tissue scaffolds and also for evaluating materials from which microelectrodes will be fabricated. As with many other in vitro studies, cell adhesion and proliferation are strong indicators for cell activity associated with a material. This is particularly relevant for astrocytes due to their tendency to proliferate after CNS injury. Again, an ideal material for a CNS implant will reduce astrocyte adhesion and proliferation. Rat astrocytes are commonly used in experiments (CRL-2005, American Type Culture Collection, Manassas, VA). Proteins associated with astrocyte activity, such as glial fibrillary acidic protein (GFAP) can also be labeled and imaged with immunocytochemistry techniques to evaluate cell activation.

Schwann cell activity is relevant for neural cell scaffold applications in the PNS due to their presence, but it is also relevant in the CNS due to regeneration approaches that use transplanted Schwann cells to promote CNS activity and myelinate regenerating CNS axons (Huang and Huang 2006). Oligodendrocytes and olfactory ensheathing cells, too, are known to form an endoneurial sheath for growing axons and have been identified as candidates for cell transplants. Schwann cells, which promote tissue regeneration in the PNS, and olfactory ensheathing cells have been shown to enhance nerve regeneration when transplanted into the CNS (Ramon-Cueto and Valverde 1995; Schmidt and Leach 2003). The morphology of Schwann cells has been shown to be an important factor in axon guidance.

Polymer molds of Schwann cells alone have been shown to guide growing axons (Bruder et al. 2007). This stresses the importance of topography of neural tissue scaffold and also the role of supporting cells in nerve regeneration strategies.

Co-cultures of relevant cell types can provide insight into how cells will interact with a material and also interact with each other. Once implanted into the body, the material will be exposed to a number of cell types in the local environment. These cells may compete to adhere to the material surface. A co-culture is a more physiological in vitro assay because it more accurately reproduces the conditions to which the material will be exposed. Different cell types, most notably the interaction between Schwann and neural cells, also produce biochemical cues which may influence cell behavior. Trophic factors, such as BDNF, are axon chemoattractants produced by Schwann cells (Meyer et al. 1992). Assays to investigate competitive adhesion or cell organization on a material's surface can provide insight into how the neural tissue will interact with the implant. Glial markers produced by Schwann cells have been studied with immunocytochemistry techniques (Corey et al. 2007).

3.3   Protein Assays

An intermediate step between the implantation of a biomaterial and the adhesion of cells is the adsorption of proteins. After implantation, proteins rapidly adsorb to the implant surface. The type of proteins, as well as the conformation of the adsorbed proteins, is determined by the surface energy and surface roughness of the material. To promote neural cell activity and tissue regeneration, the adsorption of laminin and collagenous proteins is desirable. A NGC should be engineered to enhance and optimize protein adsorption. To evaluate a NGC candidate material, protein adsorption from a solution of laminin can be evaluated in a number of ways. A solution of a known protein concentration may be placed on a material surface and then re-analyzed for protein concentration after some of the protein has adsorbed to the material surface and, thus, is depleted from the solution. Alternatively, proteins adsorbed to a material surface can be removed by soaking in a 1% sodium dodecyl sulfate (SDS) solution, collected, and measured using a MicroBCA assay (Pierce, Rockford, Illinois) (Woo et al. 2003). Western blot analysis can be used to evaluate levels of specific proteins of interest. The quantity of proteins adsorbed to a material surface provides some insight regarding the suitability of a material for tissue growth.

3.4   In Vivo Assays

In vivo studies regarding PNS regeneration most often rely on a rat sciatic nerve transection model. The sciatic nerve runs from the back to the lower limbs and provides motor and sensory function to the limb. Transecting this nerve eliminates motor and sensory function. Without surgical repair of the transected nerve, spontaneous regeneration is rare (Panseri et al. 2008). NGC materials have been evaluated by implanting a cuff around a transected nerve and suturing the severed ends of the nerve fiber bundle in place (Belkas et al. 2004; Hudson et al. 2000; Bunge 2001; Huang and Huang 2006). Without sacrificing the animal, motor and sensory recovery can be evaluated and characterized in a variety of ways, including a walking track test and Von-Frey test (Yannas and Hill 2004; Panseri et al. 2008). Electrophysiological data, specifically the compound motor action potential and compound sensory action potential amplitudes, reveal the degree to which muscle and digital sensory nerves have been reinnervated (Archibald et al. 1995). Functional motor and sensory recovery is the ultimate goal of any treatment and is indicative of tissue regeneration. Motor and sensory functions rarely recover without appreciable tissue regeneration. Sacrificing the animal and performing histology studies on the sciatic nerve and surrounding tissue can provide more detailed information regarding protein and cellular activity and tissue regeneration. If the NGC was implanted to bridge a tissue gap between the severed ends of the nerve fiber bundle, the cross-sectional area of the tissue growing to fill the gap can be evaluated along the length of the NGC (Williams et al. 1993). The number of nerve fibers, fiber diameter, density of axons, number of unmyelinated or myelinated axons, the thickness of myelin sheath, Schwann cell populations, collagen IV deposition, and degree of vascularization are all indicators of tissue regeneration which can be evaluated by tissue staining (Mackinnon et al. 1991; Gordon et al. 2005; Panseri et al. 2008). Immunohistochemistry techniques to evaluate reactive gliosis in the CNS include the labeling of GFAP and also vimentin, proteins expressed in reactive astrocytes (Silver and Miller 2004).

4 Next Generation Biomaterials for Nerve Regeneration

Traditional attempts at creating improved nerve graft materials have focused on altering biomaterials chemistry (such as various polymers), however, to date, this approach has not provided the ultimate nerve regeneration material. In contrast, in recent years, nanomaterials have become promising candidates for a variety of tissue engineering applications. Nanomaterials are materials with at least one dimension less than 100 nm. Due to a set of fundamental properties, nanomaterials can be engineered to interact with cells and proteins with a greater degree of specificity. First, nanomaterials have increased surface areas compared to conventional materials. The specific surface area of a given mass of nanoparticles is greater than the specific surface area of the same dose of micron-scale particles. The increased surface to volume ratio and surface area allows for a greater degree of surface interactions. Nanoparticles will have a greater surface to which proteins can adsorb. Two-dimensional nanorough surfaces have a greater functional surface area than materials with micron-scale surface features due to increased roughness in the direction perpendicular to the material surface. This overall increase in the surface area of nanomaterials leads to the second property that makes nanomaterials superior, increased surface energy. The increase in surface area exposes more functional groups at the material surface; surface energy is increased as a result. A hydrophobic or hydrophilic material, when fabricated with nanoroughness, will exhibit increased hydrophobicity or hydrophilicity properties, respectively. This can also enhance the adsorption of proteins and increase cell adhesion. Third, nanomaterials are biomimetic. A material with nanoroughness more accurately resembles native tissue. ECM proteins including laminin, fibronectin, and collagen fibrils have dimensions on the nanoscale. Though cell adhesion occurs on top of a layer of proteins adsorbed to a material's surface, tissue architecture suggests that nanomaterials provide a superior foundation for tissue regeneration. In comparison to conventional materials, a protein adsorbed onto a nanomaterial can maintain a conformation more conducive to cellular adhesion (Webster et al. 2001). The combination of nanoscale control over a material's surface roughness and surface energy allow for a more precise interaction with proteins. This is critical due to the relationship between protein conformation and cell function. Studies have shown greater neural cell (PC12) adhesion, osteoblast adhesion, and ECM protein adsorption to nanomaterials compared to conventional materials with similar chemistry (Khang et al. 2007; Webster et al. 2001, Webster and Ejiofor 2004; Webster et al. 2004). Enhanced cell activity has been attributed, in part, to greater amounts of fibronectin and vitronectin adsorbed to materials with greater surface roughness (Degasne et al. 1999). Further studies have shown that the proteins adsorbed to nanomaterial surfaces may expose more amino acid binding sequences than proteins adsorbed to conventional materials (Webster et al. 2001). Specifically, vitronectin adsorbed to nanophase alumina demonstrated more unfolding to expose amino acid binding sequences such as Arginine-Glycine-Aspartic Acid-Serine (RGDS) in comparison to vitronectin adsorbed to conventional alumina (Webster et al. 2001). Though these studies relate to proteins important to osteoblast adhesion, the same concepts of protein conformational optimization on nanomaterials are relevant to materials, proteins, and cells in the neural tissue engineering field as well.

4.1 Mechanisms of Protein/Nanomaterial Interactions

To appreciate why biomimetic nanomaterials provide superior tissue engineering surfaces, the role that the ECM plays in regulating cell function must be understood. All cells exist within an ECM, the three-dimensional network of proteins that provide structural support for a biological tissue. ECM proteins interact with cell surface receptors to regulate gene expression and cell function. Conventional biomaterials with surface features on the micron scale do not resemble the natural nanoroughness of the ECM and, thus, are not biomimetic. Laminin, a critical protein for neural tissue development, is a cruciform protein approximately 70 nm in length and width. Cellular interactions with a nanoscale biomaterial may provide a more physiologically activated cell surface receptor for improved interactions and greater nerve tissue regeneration.

Examples of the superiority of nanomaterials to conventional materials can be seen in research related to a variety of tissues. The persistence of nanomaterial research and the optimism that many feel for the future of biomaterial applications of nanotechnology can be attributed to the degree of control over fine tuning material properties to customize a material for promoting or inhibiting specific cell interactions. Bone, cartilage, and vascular tissue, among others, have been shown to respond to select nanorough surfaces with increased activity (Liu and Webster 2007). In the CNS, a majority of nanomaterial applications have addressed the need for improved neural electrode interfaces for direct tissue recordings. Applications in the PNS have mostly addressed the treatment of damaged peripheral nerve with NGCs, however, NGCs have been applied to the CNS as well. For example, Schwann-cell seeded polyacrylonitrile/ polyvinylchloride channels implanted into transected rat spinal cords have increased the number of myelinated axons in regenerating tissue (Xu et al. 1995).

4.2   Neural and Glial Cell Interactions with Nanomaterials

A number of studies have examined neural and glial cell response to nanomaterials (Table 2). Stem cell differentiation to a neuronal cell lineage and the enhanced activity and adhesion of neural cells has been observed on nanomaterials (Silva et al. 2004; Hu et al. 2004; Yim et al. 2007; Fan et al. 2002). Investigations of neural cell activity on carbon nanofibers has revealed enhanced neurite outgrowth compared to cells cultured on conventional carbon fibers (McKenzie et al. 2004). For neuronal interfacing microelectrodes, carbon nanotube coatings are desirable due to their high specific surface area and electrical conductivity. In vitro patterned islands of carbon nanotubes promoted the organization of ordered neural networks (Ben-Jacob and Hanein 2008). In vivo biodegradable PLGA NGCs consisting of submicron (>280 nm) fibers successfully permitted PNS tissue regeneration when implanted around a transected sciatic nerve (Panseri et al. 2008). Scaffolds produced by self-assembling peptide molecules of IKVAV and RADA have been shown to support neural regeneration (Silva et al. 2004; Ellis-Behnke et al. 2006). The biomimetic network created by self-assembled IKVAV, specifically, was shown to support the differentiation of cells into a neuronal lineage while minimizing the activation of astrocytes.

Reduced astroglial cell activity when cultured on carbon nanofibers (McKenzie et al. 2004) and ZnO nanoparticle/polycarbonate urethane composites (Seil and Webster 2008) have been observed. High concentrations (>10% by weight) of high energy carbon nanofibers in polycarbonate urethane significantly reduced astrocyte adhesion. Pure low surface energy carbon nanofibers significantly reduced astrocyte proliferation compared to similar carbon nanofibers with micron-scale diameters. ZnO nanoparticles at high concentrations (>10% by weight) in polyurethane significantly reduced astrocyte adhesion and proliferation compared to the pure polyurethane.

Table 2 Select studies regarding neural tissue regeneration applications of submicron and nanomaterials

Reference	Material	Surface dimensions (nm)	Findings
Corey et al. (2007)	Aligned electrospun poly-l-lactate nanofibers	524±305	Longer, aligned DRG neurites
Panseri et al. (2008)	Electrospun tubes of PLGA/PCL	279±87	Regeneration of 10 mm sciatic nerve defect
Moxon et al. (2004)	Silicon coated ceramic electrode	Not reported	Decreased astrocyte adhesion and increased PC12 neurite extension
Hu et al. (2004)	Chemically modified multiwall carbon nanotubes	Not reported	Increased hippocampal neurite outgrowth and alignment
Webster et al. (2004)	Carbon nanotube/ polycarbonate urethane composite	60	Increased PC12 neurite outgrowth and decreased astrocyte activity with increased CNT wt.%
McKenzie et al. (2004)	Carbon nanotube/ polycarbonate urethane composite	60	Reduced astrocyte activity on composites with high surface energy carbon nanofibers
Silva et al. (2004)	Self-assembled peptide amphiphile molecule nanofibers	5–8	Rapid differentiation of neural progenitor cells to neurons, decreased astrocyte activity
Yim et al. (2007)	Nanoimprinted PDMS	350	Enhanced differentiation of human mesenchymal stem cells to a neuronal lineage
Fan et al. (2002)	Nanorough silicon wafers	20–50	Increased neural cell adhesion and viability
Seil and Webster (2008)	Zinc oxide nanoparticle/ polyurethane composites	60	Decreased astrocyte activity on composites with increased ZnO nanoparticle wt.%
Yang et al. (2005)	Aligned electrospun poly(L-lactic acid) nanofibers	150–500	Alignment and enhancement of neurite outgrowth from neural stem cells
Schnell et al. (2007)	Electrospun poly ε caprolactone and collagen	541±164	Enhanced neural and Schwann cell process extension

The importance of topography on promoting neural tissue growth has been identified in terms of both nanoscale and micron-scale dimensions. At the nanoscale, a rough surface promotes select protein adsorption/bioactivity and subsequent cell adhesion. At the micron scale, channels formed in the material can direct axonal growth. Topographical guidance cues have been shown to affect neural cell function, including microgrooves which promoted neurite alignment, particularly when coated with laminin (Miller et al. 2002).

5   Other Nerve Regeneration Stimuli

5.1   Electrical Stimulation

Enhanced neural tissue regeneration with applied electrical stimulation has been documented over the past few decades (Borgens et al. 1979; McDevitt et al. 1987). In both the PNS and CNS, electrical fields have been shown to accelerate regeneration. Parameters for electrical stimulation vary greatly. Both AC and DC currents with a range of voltages have been observed to enhance neural tissue regeneration. Direct current electrical stimulation is known to enhance and direct neurite outgrowth (McCaig and Rajnicek 1991). Even brief periods of stimulation caused a sustained increase in neurite growth rate (Wood and Willitz 2006). Electrical fields as low as 10 mV/mm have been shown to guide neurite outgrowth (Rajnicek et al. 1998). The direction of neurite outgrowth was dependent on cell type and the surface on which the cells were cultured. Short-period, low-frequency electrical stimulation studies indicated that 100 μs, 3V pulses delivered at a rate of 20 Hz for periods as brief as 1 h accelerated axonal regeneration and the restoration of motor function (Al-Majed et al. 2000). An in vivo model using dogs with injured spinal cords showed that an oscillating electrical field of 500–600 mV/mm significantly accelerated the return of sensory and motor function at time points of 6 weeks and 6 months (Borgens et al. 1999). While the mechanism by which electrical stimulation enhances neural cell activity are not entirely understood, studies have shown that electrical current passed through substrates (such as electrically conducting polymer polypyrrole (PPy)) can promote fibronectin protein adsorption to enhance neural cell activity in vitro (Kotwal and Schmidt 2001). Upregulation of genes associated the neural tissue growth may help explain this phenomenon. In recent studies, an increase in cAMP was induced in DRG neurons by a brief period of electrical stimulation in a rat model (Udina et al. 2008). Elevated levels of cAMP were evident after the rat was sacrificed and the DRG neurons were plated in vitro. Cells with elevated levels of cAMP produced axons of significantly greater length than control groups. This study demonstrated the sustained effect that a brief period of electrical stimulation has on the neural cell behavior and axonal outgrowth.

5.2   Conductive and Piezoelectric Materials

Piezoelectric (Aebischer et al. 1987) materials have also been shown to enhance neural cell function without any external electrical stimulation. Conductive polymers, typically PPy, have been shown to be a permissive substrate for axonal growth in vitro and in vivo. However, studies have shown that an electrical stimulus applied to the conducting polymer was needed to enhance axon outgrowth to levels beyond that which would be expected on an inert, biocompatible polymer (Schmidt et al. 1997; Zhang et al. 2006). NGCs fabricated from an electrically conductive, biodegradable polymer of PPy and poly(D,L-lactide-co-epsilon-caprolactone) (PDLLA/CL) were comparable to nonconductive NGCs made of PDLLA/CL alone (Zhang et al. 2006). A current intensity of 1.7–8.4 µA/cm led to the greatest enhancement of neurite outgrowth on the conductive PDLLA/CL and PPy surfaces. Interestingly, the authors comment on the challenges associated with developing a power supply for the application of such an electrical stimulus. Piezoelectric materials produce a transient electrical response when they are mechanically deformed. As a result, an electrical stimulus may be possible without direct connection to a voltage source. Though an electrical stimulus is a known stimulatory cue for neural development, few studies have investigated neural tissue engineering applications of piezoelectric materials. A study investigating the piezoelectric properties of a piezoelectric PVDF polymer measured a 1,200 Hz oscillating voltage with a peak of 2.5 mV (Valentini et al. 1992). Voltage output was due to the mechanical stress caused by vibrations of the incubator shelf on which the measurements were taken. This piezoelectric response enhanced neural cell (Nb2a) differentiation and neurite outgrowth compared to non-piezoelectric controls. In another study regarding electrically active materials, a high voltage (8–24 kV) was applied to PLGA films and NGCs prior to cell culture experiments or implantation to produce a surface charge on the polymer (Bryan et al. 2004). The number of cells with neurites and the number of myelinated axons was significantly increased.

5.3 Piezoelectric Nanomaterials

The next generation of biomaterials should include combinations of the above known stimulatory cues. Conductive or piezoelectric nanomaterials may elicit superior regenerative response compared to conductive, piezoelectric, or nanoscale materials alone. To date, few studies have combined nanomaterials with such known stimulatory electrical cues.

An electrical stimulus may be incorporated into a nanomaterial scaffold by selecting a material that is piezoelectric. Ceramics, such as zinc oxide (ZnO) and silicon dioxide, both have piezoelectric properties and have been fabricated into a variety of nanostructures. Though ZnO alone is not a suitable candidate for a NGC, it may be mixed with a polymer to form a nanorough composite which is flexible and can easily be fabricated into tubular shapes (Fig. 3). The piezoelectric response of nanomaterials may be superior to the piezoelectric response of a conventional material. The piezoelectric properties of the ZnO nanoparticles in these composites may be enhanced compared to their conventional material counterparts due to the large surface to volume ratio inherent to nanoparticles (Xiang et al. 2006). Atoms in nanoscale ZnO structures are able to assume different positions, due to the free

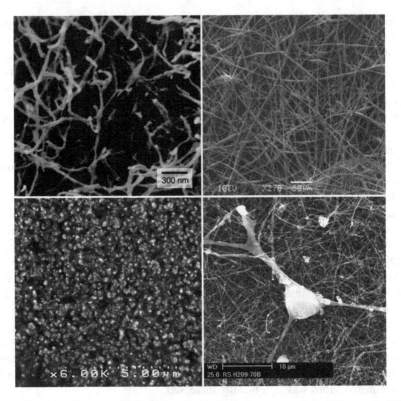

Fig. 3 *Top left* – Self assembled peptide amphiphile molecule nanofibers (Silva et al. 2004). *Top right* – electrospun PLGA/PCL fibers (Panseri et al. 2008). *Bottom left* – ZnO nanoparticles in PU (Seil and Webster 2008). *Bottom right* – neuron grown on carbon nanotube substrate (Hu et al. 2004)

boundary, which may enhance the piezoelectric effect. A significant piezoelectric response from these composites may provide the electrical stimulatory cues necessary to promote neural cell function.

A considerable number of ZnO nanostructures have been fabricated. Different structures will likely have different piezoelectric responses when mechanically deformed. Though ZnO nanoparticles with approximately spherical shapes are currently the only commercially available nanostructures (Nanophase Technologies, Romeoville, Illinois), other shapes can be synthesized with relative ease. Elongated structures may be more easily deformed to produce a piezoelectric effect.

Since a NGC is implanted to minimize tension on the nerve, it is unlikely that the scaffold will be mechanically deformed by the patient's own movement. Some form of applied stimulus will be needed to deform the scaffold. Ultrasound could potentially be applied near the area of the implanted scaffold to provide a transcutaneous stimulus. Previous studies have shown that piezoelectric materials implanted beneath 3 cm of skin and soft tissue produced an electrical response when stimulated via external ultrasound (Cochran et al. 1998). Ideally, with respect to the information gathered from studies of axon guidance under an electrical stimulus, the direction of the electric field will be controllable. This could potentially be achieved either by aligning piezoelectric nanofibers in the scaffold or by applying a directional ultrasound stimulus to randomly oriented piezoelectric nanoparticles or nanofibers.

Although ZnO and PCU composites are not biodegradable (and few, if any, piezoelectric materials are), they can serve as a useful model for investigating the efficacy of materials with both nanoscale and piezoelectric properties.

6 Conclusions and Future Directions

The most promising techniques to address neural tissue regeneration combine multiple stimuli in a single scaffold. NGCs have exhibited positive results in a number of in vivo studies and have been translated to clinical use. However, there is room for improvement in providing stimulatory cues and enhancing the interface between the NGC and the healthy tissue. Advanced properties of dynamic NGCs will deliver additional stimuli to further enhance tissue regeneration. The complexity of regeneration and the balance of minimizing inhibitory cues while providing and maximizing stimulatory cues requires multifaceted solutions. Due to the degree to which surface area can be increased and surface roughness and energy can be tailored, nanomaterials offer a promising avenue for biomaterial design. A further understanding of how even brief, low-intensity electrical stimulation can enhance axon outgrowth and neural tissue regeneration, makes the development of practical applications of electrically active materials even more appealing. Nanoscale conductive and piezoelectric materials have the potential to offer permissive environments for neural and beneficial glial cells, a reduction in the astroglial response, and critical stimulatory cues that could provide extended periods of enhanced regeneration after brief periods of treatment. Select biomaterials, including a ZnO and polymer composite with piezoelectric properties, have been shown to reduce astroglial cell activity, a critical step in the regeneration of functional neural tissue in the CNS. The piezoelectric effect of the material may provide a neural cell activity enhancing both CNS and PNS regeneration. The nanoscale dimensions of a piezoelectric material may further enhance the piezoelectric response due to the increased surface to volume ratio of the particles and the creation of a substrate that mimics the natural nanoroughness of neural tissue. Ultimately, this composite material, among other electrically active nanomaterials, when fabricated into a NGC and implanted into the body, could reduce inhibitory cues which prevent healthy tissue regeneration and provide critical stimulatory cues to promote neural cell activity and axon growth.

Acknowledgments The authors would like to thank the National Science Foundation for fellowship support (Brown University GK-12) and the Hermann Foundation for funding part of the work presented here. The authors thank Geoffrey Williams of the Leduc Bioimaging Facility at Brown University for SEM assistance for some of the work presented here.

References

Aebischer P, Valentini RF, Dario P, Domenici C, Galletti PM. Piezoelectric guidance channels enhance regeneration in the mouse sciatic nerve after axotomy. *Brain Research* 1987, 436:165–168.

Al-Majed AA, Neumann CM, Brushart TM, Gordon T. Brief electrical stimulation promotes the speed and accuracy of motor axonal regeneration. *Journal of Neuroscience* 2000, 20:2602–2608.

Archibald SJ, Shefner J, Krarup C, Madison RD. Monkey median nerve repaired by nerve graft or collagen nerve guide tube. *Journal of Neuroscience* 1995, 15:4109–4123.

Belkas JS, Shoichet MS, Midha R. Peripheral nerve regeneration through guidance tubes. *Neurological Research* 2004, 26:151–160.

Ben-Jacob E, Hanein Y. Carbon nanotube micro-electrodes for neuronal interfacing. *Journal of Material Chemistry* 2008, 18:5181–5186.

Borgens RB, Vanable JW, Jaffe LF. Small artificial current enhance Xenopus limb regeneration. *Journal of Experimental Zoology* 1979, 207:217–226.

Borgens RB, Toombs JP, Breur G, et al. An imposed oscillating electrical field improves the recovery of function in neurologically complete paraplegic dogs. *Journal of Neurotrauma* 1999, 16:639–657.

Bregman BS, McAtee M, Dai HN, Kuhn PL. Neurotrophic factors increase axonal growth after spinal cord injury and transplantation in the adult rat. *Experimental Neurology* 1997, 148:475–494.

Bruder JM, Lee AP, Hoffman-Kim D. Biomimetic materials replicating Schwann cell topography enhance neuronal adhesion and neurite alignment in vitro. *Journal of Biomaterial Science Polymer Edition* 2007, 18:967–982.

Bryan DJ, Tang JB, Doherty SA, et al. Enhanced peripheral nerve regeneration through a poled bioresorbable poly(lactic-co-glycolic acid) guidance channel. *Journal of Neural Engineering* 2004, 1:91–98.

Bunge MB. Book review: bridging areas of injury in the spinal cord. *Neuroscientist* 2001, 7:325–339.

Cochran VB, Kadaba MP, Palmieri VR. External ultrasound can generate microampere direct currents in vivo from implanted piezoelectric materials. *Journal of Orthopaedic Research* 1998, 6:145–147.

Corey JM, Lin DY, Mycek KB, Chen Q, et al. Aligned electrospun nanofibers specify the direction of dorsal root ganglia neurite growth. *Journal of Biomedical Materials Research Part A* 2007, 83A:636–645.

Degasne I, Basle MF, Demais V, et al. Effects of roughness, fibronectin and vitronectin on attachment, spreading, and proliferation of human osteoblast-like cells (saos-2) on titanium surfaces. *Calcified Tissue International* 1999, 64:499–507.

Ellis-Behnke RG, Liang YX, You SW, et al. Nano neuro knitting: peptide nanofiber scaffold for brain repair and axon regeneration with functional return of vision. *Proceedings of the National Academy of Science USA* 2006, 103:5054–5059.

Fan YW, Cui FZ, Hou SP, Xu QY, Chen LN, Lee IS. Culture of neural cells on silicon wafers with nano-scale surface topography. *Journal of Neuroscience Methods* 2002, 120:17–23.

Gordon T, Boyd JG, Sulaiman OAR. Experimental approaches to promote functional recovery after severe peripheral nerve injuries. *European Surgery* 2005, 37:193–203.

Henderson CE, Phillips HS, Pollock RA. GDNF: a potent survival factor for motoneurons present in peripheral nerve and muscle. *Science* 1994, 266:1062–1064.

Hu H, Ni Y, Montana V, Haddon RC, Parpura V. Chemically functionalized carbon nanotubes as substrates for neuronal growth. *Nano letters* 2004, 4:507–511.

Huang YC, Huang YY. Biomaterials and strategies for nerve regeneration. *Artificial Organs* 2006, 30:514–522.

Hudson TW, Evans GRD, Schmidt CE. Engineering strategies for peripheral nerve repair. *Orthopedic Clinics of North America* 2000, 31:485–497.

Khang D, Sung Yeol K, Liu-Snyder P, Palmore GTR, Durbin SM, Webster TJ. Enhanced fibronectin adsorption on carbon nanotube/poly(carbonate) urethane: independent role of surface nanoroughness and associated surface energy. *Biomaterials* 2007, 28:4756–4768.

Kotwal A, Schmidt CE. Electrical stimulation alters protein adsorption and nerve cell interactions with electrically conducting biomaterials. *Biomaterials* 2001, 22:1055–1064.

Liu H, Webster TJ. Nanomedicine for implants: a review of studies and necessary experimental tools. *Biomaterials* 2007, 28:354–369.

Mackinnon SE, Dellon AL, Obrien JP. Changes in nerve fiber numbers distal to a nerve repair in the rat sciatic nerve model. *Muscle and Nerve* 1991, 14:1116–1122.

McCaig CD, Rajnicek AM. Electrical fields, nerve growth and nerve regeneration. *Experimental Physiology* 1991, 76:473–494.

McDevitt L, Fortner P, Pomeranz B. Application of a weak electric field to the hindpaw enhances sciatic motor nerve regeneration in the adult rat. *Brain Research* 1987, 416:308–314.

McKenzie JL, Waid MC, Shi R, Webster TJ. Decreased functions of astrocytes on carbon nanofiber materials. *Biomaterials* 2004, 25:1309–1317.

Meek MF, Coert JH. US Food and Drug Administration/Conformit Europe-approved absorbable nerve conduits for clinical repair of peripheral and cranial nerves. *Annals of Plastic Surgery* 2008, 60:466–472.

Meyer M, Matsuoka I, Wetmore C, Olson L, Thoenen H. Enhanced synthesis of brain-derived neurotrophic factor in the lesioned peripheral nerve: different mechanisms are responsible for the regulation of BDNF and NGF mRNA. *Journal of Cell Biology* 1992, 119:45–54.

Miller C, Jeftinija S, Mallapragada S. Synergistic effects of physical and chemical guidance cues on neurite alignment and outgrowth on biodegradable polymer substrates. *Tissue Engineering* 2002, 8:367–378.

Moxon KA, Kalkhoran NM, Markert M, Sambito MA, McKenzie JL, Webster TJ. Nanostructured surface modification of ceramic-based microelectrodes to enhance biocompatibility for a direct brain-machine interface. *IEEE Transactions on Bio-Medical Engineering* 2004, 51:881–889.

National Spinal Cord Injury Statistical Center, University of Alabama at Birmingham, 2007 Annual Statistical Report, December, 2007.

Panseri S, Cunha C, Lowery J, et al. Electrospun micro-and nanofiber tubes for functional nervous regeneration in sciatic nerve transections. *BMC Biotechnology* 2008, 8:39–51.

Polikov VS, Tresco PA, Reichert WM. Response of brain tissue to chronically implanted neural electrodes. *Journal of Neuroscience Methods* 2005, 148:1–18.

Rajnicek AM, Robinson KR, McCaig CD. The direction of neurite growth in a weak DC electric field depends on the substratum: contributions of adhesivity and net surface charge. *Developmental Biology* 1998, 203:412–423.

Ramon-Cueto A, Valverde A. Olfactory bulb ensheathing glia: a unique cell type with axonal growth-promoting properties. *Glia* 1995, 14:163–173.

Schmidt CE, Leach JB. Neural tissue engineering: strategies for repair and regeneration. *Annual Review of Biomedical Engineering* 2003, 5:293–347.

Schmidt CE, Shastri VR, Vacanti JP, Langer R. Stimulation of neurite outgrowth using an electrically conducting polymer. *Proceedings of the National Academy of Sciences of the United States of America* 1997, 94:8948–8953.

Schnell E, Klinkhammer K, Balzer S, Brook G, Klee D, Dalton P, Mey J. Guidance of glial cell migration and axonal growth on electrospun nanofibers of poly-ε-caprolactone and a collagen/poly-ε-caprolactone blend. *Biomaterials* 2007, 28:3012–3025.

Seil JS, Webster TJ. Decreased astroglial cell adhesion and proliferation on zinc oxide nanoparticle polyurethane composites. *International Journal of Nanomedicine* 2008, 3:523–531.

Silva GA, Czeisler C, Niece KL, et al. Selective differentiation of neural progenitor cells by high-epitope density nanofibers. *Science* 2004, 303:1352–1355.

Silver J, Miller JH. Regeneration beyond the glial scar. *Nature Reviews: Neuroscience* 2004, 5:146–156.

Udina E, Furey M, Busch, S, Silver J, Gordon T, Fouad, K. Electrical stimulation of intact peripheral sensory axons in rats promotes outgrowth of their central projections. *Experimental Neurology* 2008, 210:238–247.

Valentini RF, Vargo TG, Gardella JA, Aebischer P. Electrically charged polymeric substrates enhance nerve fibre outgrowth in vitro. *Biomaterials* 1992, 13:183–190.

Wasielewski RC, Crossett LS, Rubash HE. Neural and vascular injury in total hip arthroplasty. *Orthopedic Clinics of North America* 1992, 23:219–235.

Webster TJ, Ejiofor JU. Increased osteoblast adhesion on nanophase metals: Ti, Ti6Al4V, and CoCrMo. *Biomaterials* 2004, 25:4731–4739.

Webster TJ, Schadler LS, Siegel RW, Bizios R. Mechanisms of enhanced osteoblast adhesion on nanophase alumina involve vitronectin. *Tissue Engineering* 2001, 7:291–301.

Webster TJ, Waid MC, McKenzie JL, Price RL, Ejiofor JU. Nano-biotechnology: carbon nanofibres as improved neural and orthopaedic implants. *Nanotechnology* 2004, 15:48–54.

Williams LR, Longo FM, Powell HC, Lundborg G, Varon S. Spatial-temporal progress of peripheral nerve regeneration within a silicone chamber: parameters for a bioassay. *Journal of Comparative Neurology* 1993, 218:460–470.

Woo KM, Chen VJ, Ma PX. Nano-fibrous scaffolding architecture selectively enhances protein adsorption contributing to cell attachment. *Journal Biomaterials Research Part A* 2003, 67A:531–537.

Wood M, Willitz RK. Short-duration, DC electrical stimulation increases chick embryo DRG neurite outgrowth. *Bioelectromagnetics* 2006, 27:328–331.

Xiang HJ, Yang J, Hou JG, Zhu Q. Piezoelectricity in ZnO nanowires: a first-principles study. *Applied Physics Letters* 2006, 89:223111–223113.

Xu XM, Guenard V, Kleitman N, Bunge MB. Axonal regeneration into Schwann cell-seeded guidance channels grafted into transected adult rat spinal cord. *Journal of Comparative Neurology* 1995, 351:145–160.

Yang F, Murugan R, Wang S, Ramakrishna S. Electrospinning of nano/micro scale poly(L-lactic acid) aligned fibers and their potential in neural tissue engineering. *Biomaterials* 2005, 26:2603–2610.

Yannas IV, Hill BJ. Selection of biomaterials for peripheral nerve regeneration using data from the nerve chamber model. *Biomaterials* 2004, 25:1593–1600.

Yim EKF, Pang SW, Leong KW. Synthetic nanostructures inducing differentiation of human mesenchymal stem cells into neuronal lineage. *Experimental Cell Research* 2007, 313:1820–1829.

Zhang Z, Rouabhia M, Wang Z, et al. Electrically conductive biodegradable polymer composite for nerve regeneration: Electricity-stimulated neurite outgrowth and axon regeneration. *Artificial Organs* 2006, 31:13–22.

Chapter 6
Biodegradable Metals and Responsive Biosensors for Musculoskeletal Applications

Huinan Liu

Abstract Expenditures in musculoskeletal injuries and diseases are projected to continuously increase due to the aging population and the change of lifestyles. The development of effective biomaterials and novel devices for musculoskeletal applications are one of the most important tasks of biomedical research. This chapter will introduce hard and soft tissues in the human musculoskeletal system. The physiochemical and mechanical properties of natural bone and ligaments will be reviewed so as to design novel biomaterials and devices to repair or regenerate these tissues. Biomaterials, including metals, ceramics and polymers, used for the repair and regeneration of bone and ligament tissues will be discussed. Recent research efforts and renewed interests in developing resorbable metallic biomaterials such as magnesium (Mg) alloys for orthopedic and craniofacial applications will be highlighted. Lastly, growing clinical interests in the development of responsive biosensors for orthopedic/craniofacial applications will be discussed.

Keywords Biomaterials • Biodegradable metals • Magnesium alloys • Biosensors • Musculoskeletal applications • Orthopedic applications

1 Introduction

Musculoskeletal injuries and diseases affect hundreds of millions of people around the world, and this figure is projected to escalate in the next 10–20 years due to the aging population and sedentary lifestyles (American Academy of Orthopaedic Surgeons 2008). In the United States, musculoskeletal injuries and diseases are the

H. Liu (✉)
Department of Bioengineering, University of California Riverside, 900 University Avenue, Riverside, CA 92521, USA
e-mail: huinanliu@engr.ucr.edu

T.J. Webster (ed.), *Nanotechnology Enabled In situ Sensors for Monitoring Health*, DOI 10.1007/978-1-4419-7291-0_6, © Springer Science+Business Media, LLC 2011 leading causes of disability and visits to physicians' offices accounting for more than three million hospitalizations annually, according to the National Health Interview Survey (Weinstein 2000; Praemer et al. 1995a, b). More than 1 in 4 Americans have a musculoskeletal condition requiring medical attention. Sprains, dislocations, and fractures accounted for almost one half (46%) of all musculoskeletal injuries (Praemer et al. 1995a). The economic impact of these conditions is continuously rising. In 2004, the sum of the direct expenditures in musculoskeletal-related health care and the indirect expenditures in lost wages was estimated to be $849 billion, or 7.7% of the national gross domestic product (GDP) (American Academy of Orthopaedic Surgeons 2008). In contrast, expenditures for research in these areas currently take less than 2% of the total National Institute of Health (NIH) budget of 32 billion (Weinstein 2000). More research is urgently needed if health and economic burdens are to be contained or reduced. On January 13, 2000, the World Health Organization (WHO) formally launched 2000–2010 as the Bone and Joint Decade, a global campaign to improve the quality of life for people who have musculoskeletal conditions and to advance the understanding and treatment of these conditions through research, prevention, and education. The aims of the campaign are to raise awareness of the increasing societal impact of musculoskeletal injuries and disorders, to empower patients to participate in decisions about their care, to increase funding for prevention activities and research, and to promote cost-effective prevention and treatment (Weinstein 2000).

The development of effective treatments and associated technologies for musculoskeletal injuries and diseases is one of the most important long-term goals of health care and research. There are enormous needs to develop safer, more effective methods and therapies to stimulate the healing and regeneration of musculoskeletal tissues, including bone, cartilage, ligaments, and tendons. Bioengineers, biologists, and clinicians need to work together closely as a multidisciplinary team to develop novel tissue-like biomaterials and design innovative devices to adequately address the clinical problems. This chapter will first introduce hard and soft tissues in the human musculoskeletal system. The physiochemical and mechanical properties of natural bone and ligaments will be discussed so as to closely mimic or match their composition, microstructure, nanostructure, and properties using novel biomaterials. Biomaterials, including synthetic and natural materials, for the repair and regeneration of bone and ligament tissues will be reviewed. Recent research efforts and renewed interests in developing resorbable metallic biomaterials such as magnesium (Mg) alloys for orthopedic and craniofacial applications will be highlighted. Lastly, growing clinical interests in the development of responsive biosensors for orthopedic/craniofacial applications will also be discussed.

2 Musculoskeletal System

The musculoskeletal system is an organ system that consists of bones, muscles, tendons, ligaments, and other connective tissues to provide form, support, stability, and mobility to the body. The skeletal portion made up of bones serves both a structural function (providing support and protection for vital organs) and a reservoir function (e.g., as the storage for essential minerals). Different bones are often connected by ligaments to form a joint to allow motion. A ligament connects bone to bone and provides the stability to synovial or diarthrodial joints. In order to develop better biomaterials and novel devices for musculoskeletal tissue regeneration, it is crucial to understand the biology, physiology, and functions of these tissues. Therefore, two of the most important and frequently injured tissues, bone and ligaments, will be highlighted here.

2.1 Properties and Functions of Bone

Natural bone is a composite material composed of organic compounds (mainly collagen) reinforced with inorganic compounds (minerals). Apparently, the single mineral phase of bone is too brittle and easy to break while the single collagen phase is too soft and does not have mechanical stability (such as compression strength). The composite chemistry of bone provides both strength and resilience so that the skeleton can absorb energy when stressed without breaking. The detailed composition of bone differs depending on species, age, dietary history, health status, and anatomical location. In general, however, the inorganic phase accounts for about 70% of the dry weight of bone, and the organic matrix makes up the remainder (Buckwalter et al. 2000).

The inorganic or mineral component of bone is primarily rod-like (20–80 nm long and 2–5 nm in diameter) crystalline hydroxyapatite, $Ca_{10}(PO_4)_6(OH)_2$ or HA. Small amounts of impurities, which affect cellular functions, may be present in the mineralized HA matrix; for example, magnesium, strontium, sodium, or potassium ions may replace calcium ions, carbonate may replace phosphate groups, whereas chloride and fluoride may replace hydroxyl groups. Because the release of ions from the mineral phase of the bone matrix controls cell-mediated functions, the presence of impurities may alter certain physical properties of bone (such as solubility) and consequently important biological aspects, which are critical to normal bone function. For example, magnesium present in the mineralized matrix may enhance cellular activity and promote the growth of HA crystals and subsequent new bone formation (Buckwalter et al. 2000).

Approximately 90% of the organic phase of bone is Type I collagen; the remaining 10% consists of noncollagenous proteins and ground substances. Type I collagen found in bone is synthesized by osteoblasts and is secreted as triple helical procollagen into the extracellular matrix (ECM), where collagen molecules are stabilized by cross-linking of reactive aldehydes among the collagen chains. Generally, each of the 12 types of collagen found in the body consists of three polypeptide chains composed of approximately 1,000 amino acids each. Specifically, Type I collagen (molecular weight 139,000 Da) possesses two identical $\alpha 1(I)$ chains and one unique $\alpha 2$ chain; this configuration produces a fairly rigid linear molecule that is 300 nm long. The linear molecules (or fibers) of Type I collagen are grouped into triple helix bundles having a periodicity of 67 nm, with gaps (called hole-zones) between the ends of the molecules and pores between the sides of parallel molecules (Vigier et al. 2010; Zhu et al. 2005; Miyahara et al. 1984). The collagen fibers provide the framework and architecture of bone, with the HA particles located between the fibers. Noncollagenous proteins, for example, growth factors and cytokines (such as insulin-like growth factors and osteogenic proteins), bone inductive proteins (such as osteonectin, osteopontin, and osteocalcin), and ECM compounds (such as bone sialoprotein, bone proteoglycans, and other phospho-proteins as well as proteolipids), provide minor contributions to the overall weight of bone but provide large contributions to its biological functions. During new bone formation, noncollagenous proteins are synthesized by osteoblasts, and mineral ions (such as calcium and phosphate) are deposited into the hole-zones and pores of the collagen matrix to promote HA crystal growth. The ground substance is formed from proteins, polysaccharides, and mucopolysaccharides which acts as a cement, filling the spaces between collagen fibers and HA crystals.

It is not only the complex physiochemical properties of natural bone that make it difficult to replace, but also its dynamic ability. Bone has the capability of self-repair under excessive mechanical stresses by activating the remodeling process through the formation of a bone-modeling unit (BMU). Although the inorganic and organic components of bone have structural and some regulatory functions, the principal regulators of bone metabolism are bone cells, including osteoblasts (bone-forming cells), osteocytes (bone-maintaining cells), and osteoclasts (bone-resorbing cells). Bone as a living organ can change in size, shape, composition, microstruc-ture, and properties by its remodeling process throughout its lifetime to respond to different kinds of stress produced by physical activity or mechanical loads. The remodeling process involves the removal of old bone and regeneration of new bone at the same site.

2.2 Properties and Functions of Ligaments

Ligaments appear as dense, white, hypovascular bands of tissue and vary in size, shape, and orientation depending on their anatomical location and function. For example, medial collateral ligaments (MCLs) of the knee are broad or ribbon-shaped structures while anterior cruciate ligaments (ACLs) of the knee are cylindrical-shaped structures in the midsubstance that expand into the femur and tibia. The ACL consists of two major bundles (anterior medial [AM] and posterior lateral [PL] bundles) that merge together to provide functional heterogeneity under knee flexion and extension. ACL AM and PL bundles have unique tensioning patterns with knee motion (Harner et al. 1999). Ligaments are inserted into the bone in a complex way with region-dependent mechanical inhomogeneity and mineral distribution (Moffat et al. 2008). It has long been postulated that the controlled matrix heterogeneity inherent at the ligament-to-bone interface serves to minimize the formation of stress concentrations, and to enable the transfer of complex loads between soft tissue and bone. Insertion sites also provide blood supply to the ligament.

Ligaments are composed of fibroblasts surrounded by ECM proteins. The ECM of ligaments typically contains 60–80% water. The approximate two thirds of water are crucial for cellular function and viscoelastic behavior of ligaments. The ECM typically accounts for 20–40% of the solid portion of the ligament. The ECM of ligaments are composed mainly of collagen which accounts for approximately 70–80% of its dry weight. Type I collagen accounts for 85–90% of the collagen and the rest is composed of collagen types III, VI, V, XI, and XIV. The other 20–30% of the ECM are ground substances, including proteoglycans (<1%), elastin, and other proteins and glycoproteins such as actin, laminin, and integrins (Bray et al. 2005; Frank 2004). Not all of these components are fully understood and are still an active area of research. At the molecular level, collagen is synthesized as procollagen molecules that are secreted into the extracellular space. Once outside the cell, a post-transitional modification catalyzed by a specialized enzyme called lysyl oxidase takes place. Lysyl oxidase promotes stable crosslink formation within and between procollagen molecules. The triple helical collagen molecules then line up and begin to form fibrils and subsequently fibers. Crosslinking is critical for the high strength of collagen fibers. During growth and development, crosslinks are relatively immature and soluble but with age they mature and become insoluble and increase in strength (Frank 2004). This process is important for understanding the healing process of ligaments.

Fibroblasts are relatively few in number compared to the extracellular matrix in the total volume of ligaments. Although these cells may appear physically and functionally isolated in the ligament tissue, recent evidence has indicated that normal ligament cells may communicate by means of prominent cytoplasmic extensions (Benjamin and Ralphs 2000). Gap junctions have also been detected contributing to cell connections and in linking complex networks of cellular processes (Chi et al. 2005; Lo et al. 2002a). Cell-to-cell communication raises the possibility and the potential to coordinate cellular and metabolic responses throughout the tissue. Ligaments have three dimensional (3D) anisotropic structures. Ligament fibroblasts and collagen fibers are blended within a fascicle and are organized into parallel rows that run along the long axis of the ligament (Lo et al. 2002a, b). The cells are spindle-shaped and have long cytoplasmic projections connected by gap junctions. A complex interconnected array of cells (termed cellular matrix) extends from one end of the ligament to the other and is believed to assist communication throughout the ligament and coordinate responses to both biochemical and biomechanical signals (Bray et al. 2005; Chi et al. 2005).

One of the main functions of ligaments is to stabilize joints and help guide them through the normal range of motion but limiting excessive bone displacements under high loads. Ligaments are subjected to tensile loads as they transfer load from bone to bone along the longitudinal direction of the ligament. Thus, their biomechanical properties are usually studied via a uniaxial (along the longitudinal direction) tensile test of a bone-ligament-bone complex. The effects of age on mechanical properties of a ligament vary from ligament to ligament. For example, in the rabbit model, the stiffness and ultimate load of the femur-MCL-tibia complex (FMTC) dramatically increase during skeletal maturation from 6 to 12 months of age (Woo et al. 1986,

1990). This corresponds with a change in failure mode from the tibial insertion to the midsubstance reflecting closure of the tibial epiphysis during maturation (Woo et al. 1986). However, there was no significant change from 1 to 4 years. In contrast, the human femur-ACL-tibia complex (FATC) demonstrated a rapid decrease in the stiffness and ultimate load with increasing age (Woo et al. 1991). A rapid reduction in ligament properties and mass could occur following immediate injury (Woo et al. 1986, 1987, 1990, 1991; Newton et al. 1990; Noyes 1977).

2.3 Need for Novel Biomaterials and Devices

Bone and ligaments are the most frequently injured tissues in the musculoskeletal system, and their injuries often lead to the change of activity level, work productivity, and lifestyle or even disability in the long term. For example, bone fracture is a very common type of personal injury and can be a result of traumatic accidents, pathological diseases (such as osteoporosis) or both. As ligaments are important for limiting excessive displacements between the bones at high external loads, they are frequently injured in sports, high strenuous exercises, and accidents. The ACL and MCL of the knee are especially susceptible to injuries. As much as 90% of knee ligament injuries involve these two ligaments. In most cases, the treatment for bone fractures and ligament injuries involves a large variety of medical devices and implants that are made from different biomaterials (such as metals, ceramics, polymers, and composites). Developing better biomaterials and devices are continuous research efforts to address clinical problems and challenges.

The selection of the most appropriate material to produce an orthopedic implant or device is a very important step toward the construction of a successful product. As mentioned, the properties of constituent materials will determine, to a great extent, the properties of the final implant. So far, a wide variety of natural and synthetic biomaterials, such as polymers, ceramics, and a combination of them, have been studied for orthopedic and craniofacial applications. Synthetic materials have been the material of choice for the majority of orthopedic applications. Metals and metal alloys, such as stainless steel, CoCrMo alloy, and Ti6Al4V alloy, have been the dominant materials used in orthopedic implants. Implant loosening over time is the leading cause of clinical failure, as a result of insufficient juxtaposed bone growth. Moreover, mismatches in the mechanical properties of these permanent metallic implants and physiological bone result in "stress shielding" problems in the long term according to Wolff's law (Frost 1994; Prendergast and Huiskes 1995). That is, the implanted material shields healing bone from mechanical loading, resulting in necrosis of the surrounding bone and subsequent implant loosening. Orthopedic fixation devices such as screws are also widely used in orthopedic surgeries. A large percentage of these devices are made from metallic materials due to their excellent strength (Kurosaka et al. 1987; Shen et al. 2010; Spindler et al. 2004; Steiner et al. 1994; Eriksson et al. 2001; Kartus et al. 1999; Myers et al. 2008; Moisala et al. 2008; Jarvela et al. 2008; Laxdal et al. 2006; Kaeding et al.

2005; Drogset et al. 2005; Fink et al. 2000; Benedetto et al. 2000; McGuire et al. 1999; Barber et al. 1995). Titanium alloys offer good initial fixation strength as well as acceptable cytocompatibility properties for osteointegration (Li et al. 2003, 2010; Ho et al. 2010; Gay et al. 2009; Suchenski et al. 2010; Beevers 2003; Brand et al. 2000). However, the ferromagnetic nature of these metallic fixation devices can distort postoperative magnetic resonance imaging (MRI) (Shellock et al. 1992; Bowers et al. 2008; Beuf et al. 1996; Ludeke et al. 1985; Petersilge et al. 1996; Suh et al. 1998). Also, due to their permanency, if subsequent surgeries are needed, e.g., revision surgeries, these devices often complicate the surgeries and increase the risks for patients (Shen et al. 2010; Kamath et al. 2010).

Bioabsorbable natural and synthetic polymers have attracted increasing attention for use as orthopedic biomaterials during the last 20 years (Agrawal and Ray 2001; Athanasiou et al. 1998). For example, orthopedic screws made out of polyesters or polyester/ceramic composites have been recently developed. Polyester-based materials are advantageous as they would not distort MRI images and are bioabsorbable, obviating their removal for revision surgery (Beevers 2003; Weiler et al. 2000; Safran and Harner 1996). However, the rate of degradation has been highly variable among different polymeric screws, ranging from almost complete resorption within 1 year to a significant presence at 4 years postoperatively (Stahelin et al. 1997; Bostman et al. 1992; Walton and Cotton 2007; Radford et al. 2005; Warden et al. 2008). Further, breakage during implantation occurs in 5–10% of surgeries (McGuire et al. 1999; Barber et al. 1995; Lembeck and Wulker 2005; Smith et al. 2003). Other complications have been also reported, such as migration of broken screws (Baums et al. 2006; Bottoni et al. 2000; Macdonald and Arneja 2003; Shafer and Simonian 2002; Sidhu and Wroble 1997; Takizawa et al. 1998) and persistent inflammation (Walton and Cotton 2007; Takizawa et al. 1998; Bostman et al. 1990; Kwak et al. 2008). A major concern is that even with successful short-term results, osteointegration has been poor, with bone tunnel widening and incomplete filling of the tunnel observed in a significant portion of patients (Myers et al. 2008; Moisala et al. 2008; Laxdal et al. 2006; Fink et al. 2000; Weiler et al. 2002; Barber and Boothby 2007). In fact, one study completed a 10-year follow-up and found that all patients had an osseous cyst in the place of the degraded screws as observed with MRI (Warden et al. 2008). To address the problem of osteointegration, bio-composite materials, composed of polyesters combined with ceramic particles (typically tricalcium phosphate [TCP] or hydroxyapatite), have been recently introduced (Macarini et al. 2008; Tecklenburg et al. 2006; Hunt and Callaghan 2008; LeGeros 2002; Tay et al. 1999; Lu et al. 2009). However, their effectiveness in terms of osteointegration has been promising but inconclusive (Warden et al. 2008; Barber and Dockery 2008), and breakage during implantation still remains a significant problem for these polymer-based materials (Tecklenburg et al. 2006).

Thus, an ideal biomaterial, which captures the advantages of both metallic and polymeric materials, while minimizing the disadvantages, has yet to be developed. Such a biomaterial would possess high initial mechanical properties, like the metals, to provide good initial strength and minimize the potential for breakage during and after implantation. At the same time, the ideal material would have a controllable degradation and osteoinductive capacity, so that it is gradually replaced by the regenerated tissues. Finally, the ideal orthopedic biomaterial should have minimal artifacts in MRI scans. Therefore, a novel class of degradable, metallic biomaterials, specifically magnesium and its alloys, has attracted growing interest due to their special properties for orthopedic applications.

Recent improvements in the design and processing of magnesium (Mg) alloys offer an innovative alternative to current metallic and polymeric materials. Mg alloys have a modulus and tensile strength that are approximately four to tenfold and three to tenfold higher, respectively, in comparison with polymer-based materials (Paramsothy et al. 2010; Staiger et al. 2006; Witte 2010; Hort et al. 2010; Mantovani and Witte 2010; Witte et al. 2005). Furthermore, Mg alloys can degrade in vivo (12–26 weeks), and the rate of degradation can be controlled without compromising their initial mechanical properties (Staiger et al. 2006; Hort et al. 2010). Also, these alloys have been shown to be osteoinductive (Witte et al. 2005, 2006; Waselau et al. 2007; Pietak et al. 2008). Additionally, Mg alloys have been reported to produce little artifacts in MRI scans, similar to polymeric materials (Ernstberger et al. 2009, 2010). If successful, the use of Mg alloys for orthopedic implants and devices would overcome the current limitations of metallic and polymeric materials and would revolutionize orthopedic devices.

3 Biodegradable Metals: Magnesium and Magnesium Alloys

Magnesium (Mg) is an alkaline earth element with an atomic number of 12 and an atomic mass of 24.32. Pure bulk Mg has a sliver-white color after surface polishing. Magnesium occurs abundantly in nature, especially in the ocean, always in combinational forms with other elements. It is obtained chiefly by the electrolysis of fused salts containing magnesium chloride or by the thermal reduction of magnesia. Mg alloys and Mg salts have been used in automobile, aerospace, and various pharmaceuticals (such as magnesium sulfate). In recent decades, Mg alloys have attracted significant interest in medical applications, such as cardiovascular stents, orthopedic implants, and devices.

3.1 Magnesium in the Body

The body of an average 70 kg adult contains about 2,000 mEq (equivalent to 1 mol or 24.305 g) of magnesium, about 50% of which is stored in the bones, 45% existing as intracellular cations, and about 5% in the extracellular fluid (Klein et al. 1997; Caddell 1974; Harris and Wilkinson 1971). Magnesium is one of the most abundant cations in the intracellular fluids in the body (Fox et al. 2007). The level of Mg in the extracellular fluid ranges between 0.7 and

1.05 mmol/L, where homeostasis is maintained by the kidneys and intestine (Okuma 2001). Hence, Mg is an important component of human physiology. However, at serum Mg levels exceeding 1.05 mmol/L, it is known to lead to muscular paralysis, hypotension, and respiratory distress (Vormann 2003), and cardiac arrest occurs for severely high serum levels of 6–7 mmol/L. Nevertheless, the incidence of hyper-Mg is rare due to the efficient excretion in the urine (Vormann 2003; Saris et al. 2000).

Mg is one of the essential minerals besides calcium required for bone and tooth formation (Okuma 2001; Nielsen 2006; Kobayashi et al. 2002; Toba et al. 2000; Wang et al. 1994). Mg ions are essential for many enzyme activities in carbohydrates, proteins, metabolism of nucleic acids and the interaction of intracellular particles and binding of macromolecules to subcellular organelles, such as the binding of messenger ribonucleic acid (RNA) to ribosomes. Mg, a co-factor for many enzymes, is known to stabilize the structures of DNA and RNA (Saris et al. 2000). Mg ions bind to ATP and play important roles in neurochemical transmission, muscular activity, nerve conduction, and signaling (Altura and Altura 1996) (http://medical-dictionary.thefreedictionary.com/Magnesium+alloy).

Absorption of magnesium occurs in the upper small bowel by means of an active process closely related to the transport system for calcium. Magnesium is excreted mainly by the kidney. Renal excretion of magnesium increases during diuresis induced by ammonium chloride, glucose, and organic mercurials. Magnesium affects the central nervous, neuromuscular, and cardiovascular systems. Insufficient magnesium (hypomagnesemia) in the extracellular fluid increases the release of acetylcholine and can cause changes in cardiac and skeletal muscle. Some of the conditions that can produce hypomagnesemia are diarrhea, steatorrhea, chronic alcoholism, and diabetes mellitus. Hypomagnesemia may occur in newborns and infants who are fed cow's milk or artificial formulas, apparently because of the high phosphate/magnesium ratio in such diets. Hypomagnesemia is often treated with parenteral fluids containing magnesium sulfate or magnesium chloride. Excess magnesium (hypermagnesemia) in the body can slow the heartbeat, and concentrations greater than 15 mEq/L can produce cardiac arrest in diastole. Excess magnesium also causes vasodilation by direct effects on the blood vessels and by ganglionic blockade. Hypermagnesemia is usually caused by renal insufficiency and is manifested by hypotension, electrocardiographic changes, muscle weakness, sedation, and a confused mental state (Navarro-González 1998).

3.2 Potential of Using Mg Alloys in Medical Implants

Mg has a density of 1.74 g/cm^3, which is much lower than stainless steel and titanium alloys. Mg has a good damping capacity compared to widely used polymers in medical devices and a good fracture toughness compared to widely used bioceramics, such as hydroxyapatite (HAp) (Haferkamp et al. 2000). Table 1

Table 1 Mechanical properties of various implant materials in comparison to natural bone and ligaments

Properties	Bone tissue	ACL tissue	Mg alloy	Ti alloy	Stainless steel	CoCrMo	PLLA	Synthetic HAp
Density (g/cm²)	1.8–2.0	n/a	1.74–2.0	4.4–4.5	7.9–8.1	8.3–9.2	1.23–1.25	3.1
Elastic modulus (GPa)	3–20	0.1–0.5	41–45	110–117	189–205	220–230	2.2–9.5	73–117
Tensile strength (MPa)	100–200	13–46	230–250	830–950	~485	~655	16–69	n/a
Ultimate strain %	~1.5%	15–20%	6–16%	~14%	~40	~8	1–8%	n/a
Fracture toughness (MPa×m¹ᐟ²)	3–6		15–40	55–115	50–200			0.7

highlights some physical and mechanical properties of bone and ligament tissue as well as representative biomaterials which are currently used for orthopedic applications. Obviously, most nondegradable metals (Ti alloys, stainless steel, and CoCr alloys) have much higher density and mechanical properties than actual bone and ligament tissues. All these conditions generate clinical complications and necessitate additional revision surgery. In addition, metallic implants made out of Ti alloys, stainless steel, and CoCr alloys are permanent and, thus, cannot be remodeled or replaced with time with healthy bone; this results in chronic clinical problems (such as possible consistent inflammation and malnutrition of surrounding bone tissue).

Although biodegradable polymers and resorbable ceramics have been used or explored for tissue engineering and regenerative therapies, metallic biomaterials are more suitable for load-bearing applications due to their combination of strength and toughness. One limitation of current metallic biomaterials is the possible release of toxic metallic ions and/or particles through corrosion or wear (Puleo and Huh 1995; Cohen 1998; Jacobs et al. 1991, 1998a, b, 2003; Lhotka et al. 2003; Sampath 1992) that lead to inflammatory cascades that reduce biocompatibility causing metallosis and ultimately resulting in tissue loss (Granchi et al. 1999; Niki et al. 2003; Bi et al. 2001; Wang et al. 2002). Moreover, the elastic moduli are not matched with that of natural bone tissue, resulting in stress shielding effects which decreases implant stability (Nagels et al. 2003). Current metallic biomaterials are essentially neutral in vivo, remaining as permanent fixtures, which in the case of plates, screws, and pins used to secure serious fractures must be removed by a second invasive surgical procedure after sufficient healing. Recent work has shown the promise of using magnesium (Mg) alloys as a new class of biodegradable metals for use in stents as well as for orthopedic applications (Staiger et al. 2006). Mg is exceptionally light weight with a density of 1.74 g/cm³, 2.6 and 4.5 times less dense than titanium and steel which are commonly used in current medical implants and devices. Mg has greater fracture toughness than ceramic biomaterials such as hydroxyapatite (HAp), while its elastic modulus and strength are closer to those of natural tissue than other commonly used metallic implants (Table 1). These attributes make it an ideal candidate for bone scaffolds, fixation devices, and implant applications.

The low corrosion resistance of Mg, especially in electrolytic and aqueous environments, while a major drawback in engineering applications, becomes useful for biomedical applications, where the in vivo corrosion of the Mg-based implant involves the formation of a soluble, nontoxic oxide that is harmlessly excreted in the urine. Research has shown that Mg alloys can truly degrade in vivo (Staiger et al. 2006; Hort et al. 2010). For example, it has been reported that all magnesium alloys (AZ31, AZ91, WE43, LAE442) implanted into the femur of guinea pigs degraded while the polymer (SR-PLA96) showed very little degradation 18 weeks post surgery. Theoretically, the rate of degradation of Mg alloys can be controlled without compromising their initial mechanical properties. Mg alloys have also been reported to produce less artifacts than titanium in MRI scans and to have almost identical artifacting behavior as polymeric materials (Ernstberger et al. 2009, 2010).

Moreover, due to its functional roles and presence in bone tissue, Mg may actually have stimulatory effects on the growth of new bone tissue (Witte et al. 2005; Revell et al. 2004; Zreiqat et al. 2001, 2002). Mg can remain in the body and maintain mechanical integrity over a time scale of 12–18 weeks while the bone tissue heals, eventually being replaced by natural tissue (Witte et al. 2005). Mg alloys have been proven to be osteoinductive (Witte et al. 2005, 2006; Waselau et al. 2007; Pietak et al. 2008). Mg-based substrates (AZ21) support the adhesion, differentiation, and growth of stromal cells toward an osteoblast-like phenotype with the subsequent production of a bone-like matrix under in vitro conditions (Pietak et al. 2008). It has also been shown that even fast degrading magnesium scaffolds (AZ91D) have good biocompatibility properties and react in vivo with an appropriate inflammatory host response (Witte et al. 2007a). Enhanced formation of unmineralized ECM and an enhanced mineral apposition rate adjacent to the degrading magnesium scaffolds (AZ91D) were accompanied by an increased osteoclastic bone surface, which resulted in higher bone mass and a tendency for a more mature trabecular bone structure around the Mg scaffolds in a rabbit model (Witte et al. 2007b). After 6 and 18 weeks of implantation of magnesium alloys (AZ31, AZ91, WE43, and LAE442) and a degradable polymer (SR-PLA96) intramedullary into guinea pig femora, the mineralized area and trabecular apposition rate on uncalcified sections were significantly higher around Mg alloys (AZ31, AZ91, WE43, and LAE442) compared to a polymer (SR-PLA96) (Witte et al. 2005). No statistical difference in bone formation was found among these Mg alloys (AZ31, AZ91, WE43, and LAE442).

Thus, the use of Mg alloys as biomaterials in orthopedic/craniofacial implants would overcome several of the current limitations of metallic and polyester implants, namely reducing the chance of implant breakage during implantation compared to degradable polymers, promoting bone tissue regeneration, limiting MRI interference, and degrading at a controlled rate.

3.3 Challenges in the Development of Mg Alloys for Biomedical Applications

The present challenge of Mg alloy implants is in controlling their degradation rate in the physiological environment (Gu et al. 2010; Wang et al. 2010a, b; Janning et al. 2010; Song et al. 2010; Witte et al. 2010; Zheng et al. 2010; Lorenz et al. 2009; Zhang et al. 2009, 2010; Xin et al. 2008; Kannan and Raman 2008; Li et al. 2008; Levesque et al. 2008; Xu et al. 2008). In the body, Mg ions dissolve into solution during corrosion and bind to other ions before precipitating back onto the metal surface to form a reaction layer. Chlorides bind with Mg to form $MgCl_2$, which is highly soluble in aqueous solutions, leaving no resistance to further corrosion. In high pH solutions, $Mg(OH)_2$ can form on the surface creating a passivation layer protecting Mg from further corrosion (Lorenz et al. 2009). It has also been suggested that Mg-containing phosphates contribute to the good osteoconductivity of Mg alloys (Zhang et al. 2009; Ye et al. 2010; Witte et al. 2007c). Phosphate ions also play a role in protecting the corrosion by controlling the surface pH by regulating the hydrogen ion concentration. Proteins in solution also influence corrosion resistance. In addition to being dependent upon local pH and ions, the corrosion rate is also dependent on the metals that Mg is alloyed with. Manganese (Mn) and zinc (Zn) are biocompatible; aluminum has lower biocompatibility. All these alloys improve corrosion resistance, particularly when Ca^{2+} ions are present in the solution creating an insoluble protective phosphate layer (Song et al. 2010). Protective corrosion layers are generally not uniform and do not protect the surface in all areas. Corrosion also depends on the location of the implant (compact bone vs. cancellous bone) and the local environment (Witte et al. 2006). For moderate corrosion rates, excessive Mg cations are efficiently filtered by the kidneys and passed through the urine. Corrosion of Mg also leads to hydrogen gas evolution that can cause necrosis due to a separation of tissue layers, extended healing time, or obstructing the blood stream (Witte 2010; Hort et al. 2010; Witte et al. 2005, 2006). Hydrogen evolution is also associated with crack formation on the surface of the Mg, which reduces the strength of the alloy. Hydrogen evolution is directly related to the rate of corrosion. Therefore, if the corrosion rate can be controlled, the rate of hydrogen evolution can be kept at a safe value.

Ideally, Mg-based implants or devices should be designed in a way that Mg alloys could safely dissolve into the blood stream while cells migrate and replace the metal with host tissue. To achieve this goal, biosensors that can monitor hydrogen generation, ion concentrations, and the integrity of implants or even control the degradation rate with a responsive loop are needed.

3.4 Surface Treatment on Mg Alloys for Controlling Biofunctionality and Biodegradation

The surface of Mg alloys can be treated with bioresorbable ceramics or polymers along with novel signaling biomolecules (such as growth factors, DNA, and proteins) or pharmaceutical agents (such as antibiotics, anti-inflammatory drugs) to control their biodegradation and biofunctionality.

3.4.1 Bioresorbable Ceramic Coatings on Mg Alloys

Surface treatment can improve the biocompatibility of Mg alloys by reducing the degradation rate and inducing better tissue-implant integration. Calcium phosphates or its derivatives (such as hydroxyapatite and TCP) have been coated on Mg alloys to control alloy degradation and tissue integration properties since Ca and P are the major components of natural bone (Song et al. 2010; Wang et al. 2010b; Cui et al. 2008; Hiromoto and Yamamoto 2009; Wen et al. 2009; Yang et al. 2008, 2009; Zhang and Wei 2009). If osteoconductive minerals quickly form on the Mg-Zn alloy at the early stage of degradation after surface modification, and the coatings remain intact for the initial 3–6 weeks, the chances for an osteointegrated interface increase after implantation. In addition, pitting corrosion of Mg is delayed in the presence of phosphate ions (Xin et al. 2008). Therefore, calcium phosphate-based coatings, especially osteoconductive minerals such as hydroxyapatite (HA) and TCP, are very useful in further promoting osteointegration around Mg implants and constructing new bone (Cao et al. 2010; Lu et al. 2004; Stewart et al. 2004; de Lavos-Valereto et al. 2002; Sun et al. 2002; Lavos-Valereto et al. 2001; Labat et al. 1999; Tsui et al. 1998). For example, Song et al. deposited three kinds of calcium phosphate coatings, including brushite (DCPD, $CaHPO_4 \cdot 2H_2O$), hydroxyapatite (HA, $Ca10(PO_4)6(OH)_2$), and fluoridated hydroxyapatite (FHA, $Ca5(PO_4)3(OH)1-xFx$) using electrodesposition on a biodegradable Mg-Zn alloy (Song et al. 2010). The results showed that these coatings decreased the degradation rate of Mg-Zn alloy, and the precipitates formed in modified, simulated biological fluid showed higher Ca/P molar ratios on the HA- and FHA-coated Mg-Zn alloys than the uncoated and DCPD-coated Mg-Zn alloy, which facilitated bone-like apatite formation. Both the HA and FHA coating could promote the nucleation of osteoconductive minerals (bone-like apatite) for 1 month. Hiromoto et al. developed a new method of direct synthesis of HAp on Mg and AZ series alloys (AZ31, AZ61, and AZ91) by immersion in a Ca-EDTA/KH_2PO_4 solution, which can maintain a sufficiently high concentration of Ca ions on the Mg surface. Highly crystallized HAp coatings were successfully produced, and HAp-coated specimens showed 10^3–10^4 times lower anodic current density than as-polished specimens in a polarization test. The results provided strong evidence that the HAp coating could minimize the corrosion of Mg alloys (Hiromoto and Yamamoto 2009). An adherent Ca-deficient HAp coating on Mg-Zn-Ca alloy was successfully deposited by a pulse electrodeposition process without any post-treatment (Wang et al. 2010b). The slow strain rate tensile (SSRT) test results showed that the ultimate tensile strength and time of fracture for the coated Mg-Zn-Ca alloy are greater than those of the uncoated one, which is beneficial in supporting fractured bone for a longer time. Thus, HAp coatings on Mg alloys are promising for orthopedic applications. MgF_2 coatings have also been reported to reduce Mg alloy corrosion in a rabbit model (Witte et al. 2010).

3.4.2 Polymer Coating and Surface Modification

A new approach to control the corrosion rate of Mg alloys is to fabricate a polymeric membrane or coating on the surface. Wong et al. sprayed a mixture of polycaprolactone (PCL) and dichloromethane (DCM) onto AZ91 Mg alloys. The pore size was controlled during the layer-by-layer spraying process. The results demonstrated that the deposition of the polymeric membrane on Mg alloys reduced the degradation rate of AZ91 and maintained the bulk mechanical properties for a longer period of time. The in vitro results indicated that these polymeric coatings have good cytocompatibility properties with osteoblasts compared to uncoated AZ91 alloys. The in vivo study suggested that the uncoated Mg alloys degrade more rapidly than that of the polymer-coated Mg alloys. Although new bone formation was found on both coated and uncoated Mg alloys, as determined by Micro-CT, higher volumes of new bone were observed on the polymer-coated Mg alloys. Histological analysis did not show any inflammation, necrosis, or hydrogen gas accumulation on the coated Mg alloys. Collectively, these data provided evidence that polymeric coatings on Mg alloys may mediate the degradation of Mg alloys (Wong et al. 2010).

4 Biosensors for Musculoskeletal Applications

4.1 Responsive Biosensors

Responsive biosensors are electromechanical devices that measure physiologically important variables, including chemicals, biological signals, and degradation products of implants, and respond in a controllable way to their own measurements. They can be designed to have a feedback mechanism to improve sensitivity or to activate a control function that affects the analyte. Achieving scientific breakthroughs at the synthetic/biological interface will require the development of a new class of multifunctional biocompatible materials and sensing devices capable of both: (1) sensing biochemical and physical phenomena such as biodegradation and cell functions at the synthetic/biological interface and (2) controlling the biochemical reaction and biodegradation rate through feedback loops. The development of responsive biosensors has a tremendous impact on treatment strategies and the health care industry.

Responsive biosensors could also integrate with current implants and serve as a novel component of "smart" implant systems. When placed in the body, biosensors are engineered to provide an electrical stimulus to facilitate the regeneration process of the tissue and trigger controlled degradation on demand when neo-tissue has advanced to an acceptable level of maturation. For example, a sensor that measures the degradation products of a biodegradable Mg implant is under development (Schulz et al. 2010). This sensor is designed to automatically adjust (slow down or speed up) the degradation rate of Mg alloys. Chemicals released in the body during degradation of an implant or device will affect the surrounding cells and tissues. Thus, controlling the degradation rate based on the cellular or tissue response can further control the functionality of an implant. Responsive biosensors can monitor molecular and cellular activities at the implant-tissue interface to ensure that the implant is safe and effective and can provide a way to control properties of implants during degradation.

Balancing the degradation rate and bone formation rate will be the key to developing biocompatible "smart" implants. Electrochemical impedance (EI) can be used to measure the kinetics of the metal's corrosion surface. The approach for controlled degradation is to use the Mg implant in the body as an electrochemical cell. In an electrochemical cell, power is supplied to stop the galvanic reaction. The Mg implant is the cathode (negative electrode) where reduction occurs. A cathodic current is impressed to inhibit corrosion of the implant. A noncorroding electrode will be used as the anode of the electrochemical cell. The controlled degradation system will comprise an in-body module that regulates the potential on the implant to slow dissolution and control hydrogen evolution. The effectiveness of controlled degradation can be enhanced if Mg alloys can be designed to have a corrosion potential closer to zero. At the corrosion potential, the current is at a minimum and corrosion is inhibited. If the corrosion potential is closer to zero, the current will have less effect on bone cells and redox reactions.

4.2 Bone Marker Biosensors

Bone marker sensing and correlation with bone regeneration have attracted much attention in recent decades (Yun et al. 2007). The development of electronic biosensors to detect bone markers in solutions has been reported (Yun et al. 2009). Briefly, a label-free immunosensor for bone maker detection in solutions outside the body has been tested. A self-assembled monolayer (SAM) of dithiodipropionic acid was deposited on a gold electrode. Then, a streptavidin biotinylated antihuman C-terminal telopeptide antibody was conjugated on the SAM. Subsequently, EI was measured with different concentrations of the C-terminal telopeptide.

Resistance to degradation in the physiological environment can be evaluated experimentally by measuring the current density at the corrosion potential in an electrochemical polarization curve. Better corrosion resistance gives lower corrosion current densities. As the implant degrades, Mg forms a partially passive coating and degradation continues. The volume of Mg decreases as bone cells form on the implant. During this process, the electrical impedance of the implant changes because of the formation of a passive layer, the decrease in the volume of Mg, and the attachment of bone cells on the implant. The passive layer and bone cells attaching to the implant will affect the sensor's electron transfer resistance. This relationship can be modeled mathematically to better describe implant degradation and bone formation processes simultaneously.

4.3 Ion-Selective Biosensors

Ion-selective biosensors are potentially useful in sensing Mg ion released based on hydrogen absorption. Hydrogen handling and elimination are important to prevent toxicity and pain from Mg-based implant degradation. For example, palladium has been studied for hydrogen sensing because it selectively absorbs hydrogen gas and forms a chemical species known as palladium hydride causing the metal lattice to expand by 3.5%. The principle of a hydrogen sensor lies in the fact that the palladium metal hydride's electrical resistance is greater than the metal's resistance. In such systems, the absorption of hydrogen is accompanied by a measurable increase in electrical resistance.

5 Summary and Future Directions

In summary, Mg alloys are promising next-generation biomaterials for musculoskeletal-related applications due to their unique mechanical and degradation properties. Further research is necessary to improve the design and processing of novel Mg alloys for applications in orthopedic/craniofacial implants and fixation devices. Controlling the degradation and metabolism of Mg in vivo is the key issue. Lastly, Mg-based implants may play a key role in the design of responsive, sensor-based, orthopedic implants.

References

Agrawal, C.M. and R.B. Ray, *Biodegradable polymeric scaffolds for musculoskeletal tissue engineering.* J Biomed Mater Res, 2001. **55**(2): p. 141–50.

Altura, B.M. and B.T. Altura, *Role of magnesium in patho-physiological processes and the clinical utility of magnesium ion selective electrodes.* Scand J Clin Lab Invest Suppl, 1996. **224**: p. 211–34.

Athanasiou, K.A., et al., *Orthopaedic applications for PLA-PGA biodegradable polymers.* Arthroscopy, 1998. **14**(7): p. 726–37.

Barber, F.A. and M.H. Boothby, *Bilok interference screws for anterior cruciate ligament reconstruction: clinical and radiographic outcomes.* Arthroscopy, 2007. **23**(5): p. 476–81.

Barber, F.A. and W.D. Dockery, *Long-term absorption of beta-tricalcium phosphate poly-L-lactic acid interference screws.* Arthroscopy, 2008. **24**(4): p. 441–7.

Barber, F.A., et al., *Preliminary results of an absorbable interference screw.* Arthroscopy, 1995. **11**(5): p. 537–48.

Baums, M.H., et al., *Intraarticular migration of a broken biodegradable interference screw after anterior cruciate ligament reconstruction.* Knee Surg Sports Traumatol Arthrosc, 2006. **14**(9): p. 865–8.

Beevers, D.J., *Metal vs bioabsorbable interference screws: initial fixation.* Proc Inst Mech Eng H, 2003. **217**(1): p. 59–75.

Benedetto, K.P., et al., *A new bioabsorbable interference screw: preliminary results of a prospective, multicenter, randomized clinical trial.* Arthroscopy, 2000. **16**(1): p. 41–8.

Benjamin, M. and J.R. Ralphs, *The cell and developmental biology of tendons and ligaments.* Int Rev Cytol, 2000. **196**: p. 85–130.

Beuf, O., et al., *Magnetic resonance imaging for the determination of magnetic susceptibility of materials.* J Magn Reson B, 1996. **112**(2): p. 111–8.

Bi, Y., et al., *Titanium particles stimulate bone resorption by inducing differentiation of murine osteoclasts.* J Bone Joint Surg Am, 2001. **83A**(4): 501–8.

Bostman, O., et al., *Foreign-body reactions to fracture fixation implants of biodegradable synthetic polymers.* J Bone Joint Surg Br, 1990. **72**(4): p. 592–6.

Bostman, O., et al., *Degradation and tissue replacement of an absorbable polyglycolide screw in the fixation of rabbit femoral osteotomies.* J Bone Joint Surg Am, 1992. **74**(7): p. 1021–31.

Bottoni, C.R., et al., *An intra-articular bioabsorbable interference screw mimicking an acute meniscal tear 8 months after an anterior cruciate ligament reconstruction.* Arthroscopy, 2000. **16**(4): p. 395–8.

Bowers, M.E., et al., *Effects of ACL interference screws on articular cartilage volume and thickness measurements with 1.5 T and 3 T MRI.* Osteoarthritis Cartilage, 2008. **16**(5): p. 572–8.

Brand, J., Jr., et al., *Graft fixation in cruciate ligament reconstruction.* Am J Sports Med, 2000. **28**(5): p. 761–74.

Bray, R.C., et al., *Normal ligament structure, physiology and function.* Sports Med Arthrosc Rev, 2005. **13**(3): p. 127–35.

Buckwalter, J.A., et al., *Orthopaedic basic science: biology and biomechanics of the musculoskeletal system.* 2nd ed. 2000. Rosemont: American Academy of Orthopaedic Surgeons. 873p.

Caddell, J., *Letter: on the role of magnesium depletion in severely malnourished children.* J Pediatr, 1974. **84**(5): p. 781–2.

Cao, N., et al., *Plasma-sprayed hydroxyapatite coating on carbon/carbon composite scaffolds for bone tissue engineering and related tests in vivo.* J Biomed Mater Res A, 2010. **92**(3): p. 1019–27.

Chi, S.S., et al., *Gap junctions of the medial collateral ligament: structure, distribution, associations and function.* J Anat, 2005. **207**(2): p. 145–54.

Cohen, J., *Current concepts review. Corrosion of metal orthopaedic implants.* J Bone Joint Surg Am, 1998. **80**(10): p. 1554.

Cui, F.Z., et al., *Calcium phosphate coating on magnesium alloy for modification of degradation behavior.* Front Mater Sci China, 2008. **2**(2): p. 143–8.

de Lavos-Valereto, I.C., et al., *Evaluation of the titanium Ti-6Al-7Nb alloy with and without plasma-sprayed hydroxyapatite coating on growth and viability of cultured osteoblast-like cells.* J Periodontol, 2002. **73**(8): p. 900–5.

Drogset, J.O., T. Grontvedt, and A. Tegnander, *Endoscopic reconstruction of the anterior cruciate ligament using bone-patellar tendon-bone grafts fixed with bioabsorbable or metal interference screws: a prospective randomized study of the clinical outcome.* Am J Sports Med, 2005. **33**(8): p. 1160–5.

Eriksson, K., et al., *A comparison of quadruple semitendinosus and patellar tendon grafts in reconstruction of the anterior cruciate ligament.* J Bone Joint Surg Br, 2001. **83**(3): p. 348–54.

Ernstberger, T., G. Buchhorn, and G. Heidrich, *Artifacts in spine magnetic resonance imaging due to different intervertebral test spacers: an in vitro evaluation of magnesium versus titanium and carbon-fiber-reinforced polymers as biomaterials.* Neuroradiology, 2009. **51**(8): p. 525–9.

Ernstberger, T., G. Buchhorn, and G. Heidrich, *Magnetic resonance imaging evaluation of intervertebral test spacers: an experimental comparison of magnesium versus titanium and carbon fiber reinforced polymers as biomaterials.* Ir J Med Sci, 2010. **179**(1): p. 107–11.

Fox, C.H., E.A. Timm, Jr., S.J. Smith, R.M. Touyz, E.G. Bush, and P.K. Wallace, *A method for measuring intracellular free magnesium concentration in platelets using flow cytometry.* Magnesium Research 2007, **20**(3): p. 200–7.

Fink, C., et al., *Bioabsorbable polyglyconate interference screw fixation in anterior cruciate ligament reconstruction: a prospective computed tomography-controlled study.* Arthroscopy, 2000. **16**(5): p. 491–8.

Frank, C.B., *Ligament structure, physiology and function.* J Musculoskelet Neuronal Interact, 2004. **4**(2): p. 199–201.

Frost, H.M., *Wolff's Law and bone's structural adaptations to mechanical usage: an overview for clinicians.* Angle Orthod, 1994. **64**(3): p. 175–88.

Gay, S., S. Arostegui, and J. Lemaitre, *Preparation and characterization of dense nanohydroxy-apatite/PLLA composites.* Mater Sci Eng C Biomim Supram Syst, 2009. **29**(1): p. 172–7.

Granchi, D., et al., *Cytokine release in mononuclear cells of patients with Co-Cr hip prosthesis.* Biomaterials, 1999. **20**(12): p. 1079–86.

Gu, X.N., et al. *Corrosion fatigue behaviors of two biomedical Mg alloys – AZ91D and WE43.* In: *Simulated Body Fluid.* Acta Biomaterials, 2010. **6**(12): p. 4605–13.

Haferkamp, H., M. Niemeyer, R. Boehm, U. Holzkkamp, C. Jaschik, and V. Kaese, *Development, processing and applications range of magnesium lithium alloys.* Mater Sci Forum, 2000. **350/351**: p. 31–41.

Harner, C.D., et al., *Quantitative analysis of human cruciate ligament insertions.* Arthroscopy, 1999. **15**(7): p. 741–9.

Harris, I. and A.W. Wilkinson, *Magnesium depletion in children.* Lancet, 1971. **2**(7727): p. 735–6.

Hiromoto, S. and A. Yamamoto, *High corrosion resistance of magnesium coated with hydroxy-apatite directly synthesized in an aqueous solution.* Electrochim Acta, 2009. **54**(27): p. 7085–93.

Ho, W.F., et al., *Structure and mechanical properties of Ti-5Cr based alloy with Mo addition.* Mater Sci Eng C Mater Biol Appl, 2010. **30**(6): p. 904–9.

Hort, N., et al., *Magnesium alloys as implant materials – principles of property design for Mg-RE alloys.* Acta Biomater, 2010. **6**(5): p. 1714–25.

Hunt, J.A. and J.T. Callaghan, *Polymer-hydroxyapatite composite versus polymer interference screws in anterior cruciate ligament reconstruction in a large animal model.* Knee Surg Sports Traumatol Arthrosc, 2008. **16**(7): p. 655–60.

Jacobs, J.J., et al., *Release and excretion of metal in patients who have a total hip-replacement component made of titanium-base alloy.* J Bone Joint Surg Am, 1991. **73**(10): p. 1475–86.

Jacobs, J.J., J.L. Gilbert, and R.M. Urban, *Corrosion of metal orthopaedic implants.* J Bone Joint Surg Am, 1998a. **80**(2): p. 268–82.

Jacobs, J.J., et al., *Metal release in patients who have had a primary total hip arthroplasty. A prospective, controlled, longitudinal study.* J Bone Joint Surg Am, 1998b. **80**(10): p. 1447–58.

Jacobs, J.J., et al., *Metal degradation products: a cause for concern in metal-metal bearings?* Clin Orthop Relat Res, 2003. **417**: p. 139–47.

Janning, C., et al., *Magnesium hydroxide temporarily enhancing osteoblast activity and decreasing the osteoclast number in peri-implant bone remodelling.* Acta Biomater, 2010. **6**(5): p. 1861–8.

Jarvela, T., et al., *Double-bundle anterior cruciate ligament reconstruction using hamstring autografts and bioabsorbable interference screw fixation: prospective, randomized, clinical study with 2-year results.* Am J Sports Med, 2008. **36**(2): p. 290–7.

Kaeding, C., et al., *A prospective randomized comparison of bioabsorbable and titanium anterior cruciate ligament interference screws.* Arthroscopy, 2005. **21**(2): p. 147–51.

Kamath, G.V., J.C. Redfern, P.E. Greis, and R.T. Burks, *Revision anterior cruciate ligament reconstruction.* Am J Sports Med, 2010.

Kannan, M.B. and R.K. Raman, *In vitro degradation and mechanical integrity of calcium-containing magnesium alloys in modified-simulated body fluid.* Biomaterials, 2008. **29**(15): p. 2306–14.

Kartus, J., et al., *Complications following arthroscopic anterior cruciate ligament reconstruction. A 2–5-year follow-up of 604 patients with special emphasis on anterior knee pain.* Knee Surg Sports Traumatol Arthrosc, 1999. **7**(1): p. 2–8.

Klein, G.L., et al., *Dysregulation of calcium homeostasis after severe burn injury in children: possible role of magnesium depletion.* J Pediatr, 1997. **131**(2): p. 246–51.

Kobayashi, M., K. Hara, and Y. Akiyama, *Changes in bone strength and bone mineral density in rats given low magnesium (Mg) diet.* Jpn J Pharmacol, 2002. **88**: p. 105.

Kurosaka, M., S. Yoshiya, and J.T. Andrish, *A biomechanical comparison of different surgical techniques of graft fixation in anterior cruciate ligament reconstruction.* Am J Sports Med, 1987. **15**(3): p. 225–9.

Kwak, J.H., et al., *Delayed intra-articular inflammatory reaction due to poly-L-lactide bioabsorbable interference screw used in anterior cruciate ligament reconstruction.* Arthroscopy, 2008. **24**(2): p. 243–6.

Labat, B., et al., *Interaction of a plasma-sprayed hydroxyapatite coating in contact with human osteoblasts and culture medium.* J Biomed Mater Res, 1999. **46**(3): p. 331–6.

Lavos-Valereto, I.C., et al., *In vitro and in vivo biocompatibility testing of Ti-6Al-7Nb alloy with and without plasma-sprayed hydroxyapatite coating.* J Biomed Mater Res, 2001. **58**(6): p. 727–33.

Laxdal, G., et al., *Biodegradable and metallic interference screws in anterior cruciate ligament reconstruction surgery using hamstring tendon grafts: prospective randomized study of radiographic results and clinical outcome.* Am J Sports Med, 2006. **34**(10): p. 1574–80.

LeGeros, R.Z., *Properties of osteoconductive biomaterials: calcium phosphates.* Clin Orthop Relat Res, 2002. **395**: p. 81–98.

Lembeck, B. and N. Wulker, *Severe cartilage damage by broken poly-L-lactic acid (PLLA) interference screw after ACL reconstruction.* Knee Surg Sports Traumatol Arthrosc, 2005. **13**(4): p. 283–6.

Levesque, J., et al., *Design of a pseudo-physiological test bench specific to the development of biodegradable metallic biomaterials.* Acta Biomater, 2008. **4**(2): p. 284–95.

Lhotka, C., et al., *Four-year study of cobalt and chromium blood levels in patients managed with two different metal-on-metal total hip replacements.* J Orthop Res, 2003. **21**(2): p. 189–95.

Li, J.C., Y. He, and Y. Inoue, *Thermal and mechanical properties of biodegradable blends of poly(L-lactic acid) and lignin.* Polym Int, 2003. **52**(6): p. 949–55.

Li, Z., et al., *The development of binary Mg-Ca alloys for use as biodegradable materials within bone.* Biomaterials, 2008. **29**(10): p. 1329–44.

Li, J., L. Zhou, and Z.C. Li, *Microstructures and mechanical properties of a new titanium alloy for surgical implant application.* Int J Miner Metall Mater, 2010. **17**(2): p. 185–91.

Lo, I.K., et al., *The cellular networks of normal ovine medial collateral and anterior cruciate ligaments are not accurately recapitulated in scar tissue.* J Anat, 2002a. **200**(Pt 3): p. 283–96.

Lo, I.K., et al., *The cellular matrix: a feature of tensile bearing dense soft connective tissues.* Histol Histopathol, 2002b. **17**(2): p. 523–37.

Lorenz, C., et al., *Effect of surface pre-treatments on biocompatibility of magnesium.* Acta Biomater, 2009. **5**(7): p. 2783–9.

Lu, Y.P., et al., *Plasma-sprayed hydroxyapatite+titania composite bond coat for hydroxyapatite coating on titanium substrate.* Biomaterials, 2004. **25**(18): p. 4393–403.

Lu, Y., et al., *Influence of hydroxyapatite-coated and growth factor-releasing interference screws on tendon-bone healing in an ovine model.* Arthroscopy, 2009. **25**(12): p. 1427–34.

Ludeke, K.M., P. Roschmann, and R. Tischler, *Susceptibility artefacts in NMR imaging.* Magn Reson Imaging, 1985. **3**(4): p. 329–43.

Macarini, L., et al., *Poly-L-lactic acid-hydroxyapatite (PLLA-HA) bioabsorbable interference screws for tibial graft fixation in anterior cruciate ligament (ACL) reconstruction surgery: MR evaluation of osteointegration and degradation features.* Radiol Med, 2008. **113**(8): p. 1185–97.

Macdonald, P. and S. Arneja, *Biodegradable screw presents as a loose intra-articular body after anterior cruciate ligament reconstruction.* Arthroscopy, 2003. **19**(6): p. E22–4.

Mantovani, D. and F. Witte, *The attraction of a lightweight metal with mechanical properties suitable for many applications brought a renewed focus on magnesium alloys in the automotive and aerospace industries.* Acta Biomater, 2010. **6**(5): p. 1679.

McGuire, D.A., F.A. Barber, B.F. Elrod, and L.E. Paulos, *Bioabsorbable interference screws for graft fixation in anterior cruciate ligament reconstruction.* Arthroscopy, 1999. **15**(5): p. 463–73.

Miyahara, M., et al., *Formation of collagen fibrils by enzymic cleavage of precursors of type I collagen in vitro.* J Biol Chem, 1984. **259**(15): p. 9891–8.

Moffat, K.L., et al., *Characterization of the structure-function relationship at the ligament-to-bone interface.* Proc Natl Acad Sci U S A, 2008. **105**(23): p. 7947–52.

Moisala, A.S., et al., *Comparison of the bioabsorbable and metal screw fixation after ACL reconstruction with a hamstring autograft in MRI and clinical outcome: a prospective randomized study.* Knee Surg Sports Traumatol Arthrosc, 2008. **16**(12): p. 1080–6.

Myers, P., et al., *Bioabsorbable versus titanium interference screws with hamstring autograft in anterior cruciate ligament reconstruction: a prospective randomized trial with 2-year follow-up.* Arthroscopy, 2008. **24**(7): p. 817–23.

Nagels, J., M. Stokdijk, and P.M. Rozing, *Stress shielding and bone resorption in shoulder arthroplasty.* J Shoulder Elbow Surg, 2003. **12**(1): p. 35–9.

Navarro-González, J.F., *Magnesium in dialysis patients: serum levels and clinical implications.* Clin Nephrol. 1998. **49**(6): p. 373–8.

Newton, P.O., et al., *Ultrastructural changes in knee ligaments following immobilization.* Matrix, 1990. **10**(5): p. 314–9.

Nielsen, F.H., *A mild magnesium deprivation affects calcium excretion but not bone strength and shape, including changes induced by nickel deprivation, in the rat.* Biol Trace Elem Res, 2006. **110**(2): p. 133–49.

Niki, Y., et al., *Metal ions induce bone-resorbing cytokine production through the redox pathway in synoviocytes and bone marrow macrophages.* Biomaterials, 2003. **24**(8): p. 1447–57.

Noyes, F.R., *Functional properties of knee ligaments and alterations induced by immobilization: a correlative biomechanical and histological study in primates.* Clin Orthop Relat Res, 1977. **123**: p. 210–42.

Okuma, T., *Magnesium and bone strength.* Nutrition, 2001. **17**(7–8): p. 679–80.

Paramsothy, M., et al., *Simultaneous enhancement of tensile/compressive strength and ductility of magnesium alloy AZ31 using carbon nanotubes.* J Nanosci Nanotechnol, 2010. **10**(2): p. 956–64.

Petersilge, C.A., et al., *Optimizing imaging parameters for MR evaluation of the spine with titanium pedicle screws.* AJR Am J Roentgenol, 1996. **166**(5): p. 1213–8.

Pietak, A., et al., *Bone-like matrix formation on magnesium and magnesium alloys.* J Mater Sci Mater Med, 2008. **19**(1): p. 407–15.

Praemer, A., S. Furner, and D.P. Rice. (Eds), National Center for Health Statistics, National Health Interview Survey, 1995. In: *Musculoskeletal Conditions in the United States.* 1999a. Rosemont: American Academy of Orthopaedic Surgeons.

Praemer, A., S. Furner, and D.P. Rice. (Eds), National Hospital Discharge Survey, 1995. In: *Musculoskeletal Conditions in the United States.* 1999b. Rosemont: American Academy of Orthopaedic Surgeons.

Prendergast, P.J. and R. Huiskes, *The biomechanics of Wolff's law: recent advances.* Ir J Med Sci, 1995. **164**(2): p. 152–4.

Puleo, D.A. and W.W. Huh, *Acute toxicity of metal ions in cultures of osteogenic cells derived from bone marrow stromal cells.* J Appl Biomater, 1995. **6**(2): p. 109–16.

Radford, M.J., et al., *The natural history of a bioabsorbable interference screw used for anterior cruciate ligament reconstruction with a 4-strand hamstring technique.* Arthroscopy, 2005. **21**(6): p. 707–10.

Revell, P.A., et al., *The effect of magnesium ions on bone bonding to hydroxyapatite coating on titanium alloy implants.* Bioceramics, 2004. **254**: p. 447–50.

Safran, M.R. and C.D. Harner, *Technical considerations of revision anterior cruciate ligament surgery.* Clin Orthop Relat Res, 1996. **325**: p. 50–64.

Sampath, S.A., *Release and excretion of metal in patients who have a total hip-replacement component made of titanium-base alloy.* J Bone Joint Surg Am, 1992. **74**(9): p. 1431–2.

Saris, N.E., et al., *Magnesium. An update on physiological, clinical and analytical aspects.* Clin Chim Acta, 2000. **294**(1–2): p. 1–26.

Schulz, M.J., et al., *Responsive biosensors for biodegradable magnesium implants.* 2010. Lake Buena Vista: ASME.

Shafer, B.L. and P.T. Simonian, *Broken poly-L-lactic acid interference screw after ligament reconstruction.* Arthroscopy, 2002. **18**(7): p. E35.

Shellock, F.G., et al., *MR imaging and metallic implants for anterior cruciate ligament reconstruction: assessment of ferromagnetism and artifact.* J Magn Reson Imaging, 1992. **2**(2): p. 225–8.

Shen, C., et al., *Bioabsorbable versus metallic interference screw fixation in anterior cruciate ligament reconstruction: a meta-analysis of randomized controlled trials.* Arthroscopy, 2010. **26**(5): p. 705–13.

Sidhu, D.S. and R.R. Wroble, *Intraarticular migration of a femoral interference fit screw. A complication of anterior cruciate ligament reconstruction.* Am J Sports Med, 1997. **25**(2): p. 268–71.

Smith, C.A., et al., *Fracture of Bilok interference screws on insertion during anterior cruciate ligament reconstruction.* Arthroscopy, 2003. **19**(9): p. E115–17.

Song, Y., et al., *Electrodeposition of Ca-P coatings on biodegradable Mg alloy: in vitro biomineralization behavior.* Acta Biomater, 2010. **6**(5): p. 1736–42.

Spindler, K.P., et al., *Anterior cruciate ligament reconstruction autograft choice: bone-tendon-bone versus hamstring: does it really matter? A systematic review.* Am J Sports Med, 2004. **32**(8): p. 1986–95.

Stahelin, A.C., et al., *Clinical degradation and biocompatibility of different bioabsorbable interference screws: a report of six cases.* Arthroscopy, 1997. **13**(2): p. 238–44.

Staiger, M.P., et al., *Magnesium and its alloys as orthopedic biomaterials: a review.* Biomaterials, 2006. **27**(9): p. 1728–34.

Steiner, M.E., et al., *Anterior cruciate ligament graft fixation – comparison of hamstring and patellar tendon grafts.* Am J Sports Med, 1994. **22**(2): p. 240–7.

Stewart, M., J.F. Welter, and V.M. Goldberg, *Effect of hydroxyapatite/tricalcium-phosphate coating on osseointegration of plasma-sprayed titanium alloy implants.* J Biomed Mater Res A, 2004. **69**(1): p. 1–10.

Suchenski, M., et al., *Material properties and composition of soft-tissue fixation.* Arthroscopy, 2010. **26**(6): p. 821–31.

Suh, J.S., et al., *Minimizing artifacts caused by metallic implants at MR imaging: experimental and clinical studies.* AJR Am J Roentgenol, 1998. **171**(5): p. 1207–13.

Sun, L., et al., *Surface characteristics and dissolution behavior of plasma-sprayed hydroxyapatite coating.* J Biomed Mater Res, 2002. **62**(2): p. 228–36.

Takizawa, T., et al., *Foreign body gonitis caused by a broken poly-L-lactic acid screw.* Arthroscopy, 1998. **14**(3): p. 329–30.

Tay, B.K., V.V. Patel, and D.S. Bradford, *Calcium sulfate- and calcium phosphate-based bone substitutes. Mimicry of the mineral phase of bone.* Orthop Clin North Am, 1999. **30**(4): p. 615–23.

Tecklenburg, K., et al., *Prospective evaluation of patellar tendon graft fixation in anterior cruciate ligament reconstruction comparing composite bioabsorbable and allograft interference screws.* Arthroscopy, 2006. **22**(9): p. 993–9.

Toba, Y., et al., *Dietary magnesium supplementation affects bone metabolism and dynamic strength of bone in ovariectomized rats.* J Nutr, 2000. **130**(2): p. 216–20.

Tsui, Y.C., C. Doyle, and T.W. Clyne, *Plasma sprayed hydroxyapatite coatings on titanium substrates. Part 2: optimisation of coating properties.* Biomaterials, 1998. **19**(22): p. 2031–43.

United States Bone and Joint Decade: The Burden of Musculoskeletal Diseases in the United States. American Academy of Orthopaedic Surgeons. 2008; Available from: http://www.boneandjointburden.org/ on August 8, 2010.

Vigier, S., et al., *Collagen supramolecular and suprafibrillar organizations on osteoblasts long-term behavior: benefits for bone healing materials.* J Biomed Mater Res A, 2010. **94**(2): p. 556–67.

Vormann, J., *Magnesium: nutrition and metabolism.* Mol Aspects Med, 2003. **24**(1–3): p. 27–37.

Walton, M. and N.J. Cotton, *Long-term in vivo degradation of poly-L-lactide (PLLA) in bone.* J Biomater Appl, 2007. **21**(4): p. 395–411.

Wang, C., C.J. Lee, and Y.L. Xiong, *Interrelationship of dietary energy, calcium and magnesium content on bone-composition and strength in oophorohysterectomized rats.* FASEB J, 1994. **8**(5): p. A707.

Wang, M.L., et al., *Titanium particles suppress expression of osteoblastic phenotype in human mesenchymal stem cells.* J Orthop Res, 2002. **20**(6): p. 1175–84.

Wang, J., et al., *Microstructure and corrosion properties of as sub-rapid solidification Mg-Zn-Y-Nd alloy in dynamic simulated body fluid for vascular stent application.* J Mater Sci Mater Med, 2010a. **21**(7): p. 2001–8.

Wang, H.X., et al., *In vitro degradation and mechanical integrity of Mg-Zn-Ca alloy coated with Ca-deficient hydroxyapatite by the pulse electrodeposition process.* Acta Biomater, 2010b. **6**(5): p. 1743–8.

Warden, W.H., D. Chooljian, and D.W. Jackson, *Ten-year magnetic resonance imaging follow-up of bioabsorbable poly-L-lactic acid interference screws after anterior cruciate ligament reconstruction.* Arthroscopy, 2008. **24**(3): 370e1–3.

Waselau, M., et al., *Effects of a magnesium adhesive cement on bone stability and healing following a metatarsal osteotomy in horses.* Am J Vet Res, 2007. **68**(4): p. 370–8.

Weiler, A., et al., *Biodegradable implants in sports medicine: the biological base.* Arthroscopy, 2000. **16**(3): p. 305–21.

Weiler, A., et al., *Tendon healing in a bone tunnel. Part II: histologic analysis after biodegradable interference fit fixation in a model of anterior cruciate ligament reconstruction in sheep.* Arthroscopy, 2002. **18**(2): p. 124–35.

Weinstein, S.L., *2000–2010: the bone and joint decade.* J Bone Joint Surg Am, 2000. **82**(1): p. 1–3.

Wen, C., et al., *Characterization and degradation behavior of AZ31 alloy surface modified by bone-like hydroxyapatite for implant applications.* Appl Surf Sci, 2009. **255**(13–14): p. 6433–8.

Witte, F., *The history of biodegradable magnesium implants: a review.* Acta Biomater, 2010. **6**(5): p. 1680–92.

Witte, F., et al., *In vivo corrosion of four magnesium alloys and the associated bone response.* Biomaterials, 2005. **26**(17): p. 3557–63.

Witte, F., et al., *In vitro and in vivo corrosion measurements of magnesium alloys.* Biomaterials, 2006. **27**(7): p. 1013–8.

Witte, F., et al., *Biodegradable magnesium scaffolds: Part 1: appropriate inflammatory response.* J Biomed Mater Res A, 2007a. **81**(3): p. 748–56.

Witte, F., et al., *Biodegradable magnesium scaffolds: Part II: peri-implant bone remodeling.* J Biomed Mater Res A, 2007b. **81**(3): p. 757–65.

Witte, F., et al., *Biodegradable magnesium-hydroxyapatite metal matrix composites.* Biomaterials, 2007c. **28**(13): p. 2163–74.

Witte, F., et al., *In vivo corrosion and corrosion protection of magnesium alloy LAE442.* Acta Biomater, 2010. **6**(5): p. 1792–9.

Wong, H.M., et al., *A biodegradable polymer-based coating to control the performance of magnesium alloy orthopaedic implants.* Biomaterials, 2010. **31**(8): p. 2084–96.

Woo, S.L., et al., *Tensile properties of the medial collateral ligament as a function of age.* J Orthop Res, 1986. **4**(2): p. 133–41.

Woo, S.L., et al., *The biomechanical and morphological changes in the medial collateral ligament of the rabbit after immobilization and remobilization.* J Bone Joint Surg Am, 1987. **69**(8): p. 1200–11.

Woo, S.L., K.J. Ohland, and J.A. Weiss, *Aging and sex-related changes in the biomechanical properties of the rabbit medial collateral ligament.* Mech Ageing Dev, 1990. **56**(2): p. 129–42.

Woo, S.L., et al., *Tensile properties of the human femur-anterior cruciate ligament-tibia complex. The effects of specimen age and orientation.* Am J Sports Med, 1991. **19**(3): p. 217–25.

Xin, Y., et al., *Influence of aggressive ions on the degradation behavior of biomedical magnesium alloy in physiological environment.* Acta Biomater, 2008. **4**(6): p. 2008–15.

Xu, L., et al., *In vitro corrosion behaviour of Mg alloys in a phosphate buffered solution for bone implant application.* J Mater Sci Mater Med, 2008. **19**(3): p. 1017–25.

Yang, J.X., et al., *Calcium phosphate coating on magnesium alloy by biomimetic method: investigation of morphology, composition and formation process*. Front Mater Sci China, 2008. **2**(2): p. 149–55.

Yang, J.X., et al., *Characterization and degradation study of calcium phosphate coating on magnesium alloy bone implant in vitro*. IEEE Trans Plasma Sci, 2009. **37**(7): 1161–68.

Ye, X., et al., *In vitro corrosion resistance and cytocompatibility of nano-hydroxyapatite reinforced Mg-Zn-Zr composites*. J Mater Sci Mater Med, 2010. **21**(4): p. 1321–8.

Yun, Y., et al., *A nanotube array immunosensor for direct electrochemical detection of antigen-antibody binding*. Sens Actuators B Chem, 2007. **123**(1): p. 177–82.

Yun, Y.H., et al., *A label-free electronic biosensor for detection of bone turnover markers*. Sensors, 2009. **9**(10): p. 7957–69.

Zhang, Y. and M. Wei, *Controlling the biodegradation rate of magnesium using sol-gel and apatite coatings*. Int J Modern Phys B, 2009. **23**(6–7): p. 1897–903.

Zhang, Y., G. Zhang, and M. Wei, *Controlling the biodegradation rate of magnesium using biomimetic apatite coating*. J Biomed Mater Res B Appl Biomater, 2009. **89B**(2): p. 408–14.

Zhang, S., et al., *Research on an Mg-Zn alloy as a degradable biomaterial*. Acta Biomater, 2010. **6**(2): p. 626–40.

Zheng, Y.F., et al., *In vitro degradation and cytotoxicity of Mg/Ca composites produced by powder metallurgy*. Acta Biomater, 2010. **6**(5): p. 1783–91.

Zhu, B., et al., *Alignment of osteoblast-like cells and cell-produced collagen matrix induced by nanogrooves*. Tissue Eng, 2005. **11**(5–6): p. 825–34.

Zreiqat, H., et al., *Mechanisms of magnesium-stimulated adhesion of osteoblastic cells to commonly used orthopaedic implants*. J Bone Miner Res, 2001. **16**: p. S328.

Zreiqat, H., et al., *Mechanisms of magnesium-stimulated adhesion of osteoblastic cells to commonly used orthopaedic implants*. J Biomed Mater Res, 2002. **62**(2): p. 175–84.

Chapter 7
Carbon Nanotube-Based Orthopedic Implant Sensors

Sirinrath Sirivisoot

Abstract In this chapter, the use of multiwalled carbon nanotubes (MWCNTs) as sensors for healthy bone growth will be discussed. MWCNTs are cytocompatible with osteoblasts (bone-forming cells) and enhance osteoblast calcium deposition, and thus, bone formation. Moreover, here, MWCNTs have been grown out of nanopores of anodized titanium (MWCNT-Ti), which has a surface layer of titanium oxide (a popular chemistry for orthopedic implants), to serve as a novel in situ orthopedic implant sensor. The electrochemical responses of MWCNT-Ti have been investigated in an attempt to ascertain if MWCNT-Ti can serve as novel in situ sensors for bone formation. MWCNT-Ti have been subjected to a ferri/ferrocyanide redox couple and its electrochemical behaviors measured. Cyclic voltammograms (CV) have demonstrated enhanced redox potentials of the MWCNT-Ti compared to Ti alone. These redox signals have been superior to that obtained by bare Ti, which does not sense either oxidation or reduction peaks in CV. Moreover, the redox reactions of MWCNT-Ti have been tested in a solution of extracellular bone components synthesized by osteoblasts in vitro. It has been found that MWCNT-Ti also exhibited well-defined and persistent CV, similar to the ferri/ferrocyanide redox reaction. The observed higher electrodic performance and electrocatalytic activity the MWCNT-Ti senses compared to bare Ti, are likely because of the fact that MWCNTs enhance direct electron transfer and facilitate double layer effects, leading to a strong redox signal. These collective results may encourage the further study and modification of MWCNT-Ti to sense new bone growth in situ next to implants and perhaps to monitor other events (such as infection and/or scar tissue formation) to improve the current clinical diagnosis of orthopedic implant success or failure.

Keywords Osteoblasts • Titanium • Multiwalled carbon nanotubes • Biosensing

S. Sirivisoot(✉)
Wake Forest Institute for Regenerative Medicine, Wake Forest University,
Health Sciences, 391 Technology Way, Winston-Salem, NC 27157, USA
e-mail: ssirivis@fubmc.edu

T.J. Webster (ed.), *Nanotechnology Enabled In situ Sensors for Monitoring Health*,
DOI 10.1007/978-1-4419-7291-0_7, © Springer Science+Business Media, LLC 2011

1 Introduction

Osteoblasts and osteoclasts are located in bone, a natural nanostructured-mineralized organic matrix. While osteoblasts make bone, osteoclasts decompose bone by releasing acid that degrades calcium phosphate-based apatite minerals into an aqueous environment. The synthesis, deposition, and mineralization of this organic matrix, in which osteoblasts proliferate and mineralize (that is, deposit calcium), require the ordered expression of a number of osteoblast genes. Bone has the ability to self-repair or remodel routinely. However, osteoporosis (unbalanced bone remodeling) and other joint diseases (such as osteoarthritis, rheumatoid arthritis, or traumatic arthritis) can lead to bone fractures. These disabilities associated with bone lead to difficulties in performing common activities and may require an orthopedic implant. However, the average functional lifetime of, for example, a hip implant (usually composed of titanium) is only 10–15 years. A lack of fixation into surrounding bone eventually loosens the implant and is the most common cause of hip replacement failure.

Approximately, 200,000 total and 100,000 partial hip replacements as well as 36,000 revision hip replacement surgeries were performed in the United States in 2003 (Zhan et al. 2007). Unfortunately, only limited techniques, such as insensitive imaging (through X-rays or bone scans), exist to determine if sufficient bone growth is occurring next to a titanium implant after surgery. Thus, techniques used in orthopedic diagnostics need to more accurately identify musculoskeletal injuries and conditions after implantation. Although there have been improvements in implants to increase bone formation, the clinical diagnosis of new bone growth surrounding implants remains problematic, sometimes significantly increasing hospital stays and decreasing the ability of doctors to quickly prescribe a change in action if new bone growth is insufficient.

Currently, a physical examination (e.g., palpation, or laboratory testing) might be completed before imaging techniques are used to inform a clinician about a patient's health. Although advanced imaging techniques, such as bone scans, computer tomography scans, and radiographs (X-rays) are important in medical diagnosis, each has its own limitations and difficulties. A bone scan is used to identify areas of abnormal active bone formation, such as in arthritis, infection, or bone cancer. However, bone scans require an injection of a radioactive substance (e.g., technetium) and a prolonged delay for absorbance before performing the scan. Computer tomography combines X-rays with computer technology to produce a two dimensional cross-sectional image of a body on the computer screen. Although this technique produces more details than an X-ray, in some cases (e.g., severe trauma to the chest, abdomen, pelvis or spinal cord), a dye (e.g., barium sulfate) must be injected in order to improve the clarity of the image. This often causes pain to the patient. Another technique, called electromyography, has been used to analyze/diagnose nerve functions inside body conditions. Thin electrodes are placed in soft tissues to help analyze and record electrical activity in the muscles. However, this electrode technique leads to pain and discomfort for the patient. When the needles are removed, soreness and bruising can occur.

Electrochemical biosensors on the implant itself show promise for medical diagnostics in situ to possibly determine new bone growth surrounding the implant. Yet, this technology has not been fully explored. Incorporation of such electrochemical biosensors into current bone implants may be possible through nanotechnology; different types of nanoscale biomaterials have varying abilities to enhance in vitro/vivo bone formation. Carbon nanotubes (CNTs) are macromolecules of carbon, classified either as single-walled carbon nanotubes (SWCNTs) with diameters of 0.4–2 nm or multiwalled carbon nanotubes (MWCNTs) with diameters of 2–100 nm (Iijima 1991; Lin et al. 2004). Due to their unique electrical, mechanical, chemical, and biological properties (Webster et al. 2004; Smart et al. 2006; Zanello 2006; Zanello et al. 2006; Balani et al. 2007; Chen et al. 2007; Wei et al. 2007; Harrison and Atala 2007), CNTs have shown promise for bone implantation. For example, nanocomposites of polylactic acid and MWCNTs increased osteoblast proliferation by 46% and calcium production by greater than 300% when an alternating current was applied in vitro (Supronowicz et al. 2002; Ciombor and Aaron 2005). In addition, combining MWCNTs with precursor powders improves the mechanical properties of as-aligned hydroxyapatite (HA) composite coatings (Chen et al. 2007). MWCNTs, reinforced with HA coatings, promoted human osteoblast proliferation in vitro, as observed in the appearance of cells near MWCNT regions (Balani et al. 2007). Osteoblasts extended in all directions within the CNT scaffold that formed on polycarbonate membranes (Aoki et al. 2005). Osteoblasts significantly enhanced their adhesion on vertically aligned MWCNT arrays, according to Giannona et al. by recognizing nanoscaled features (Giannona et al. 2007). In addition, Zanello et al. (2006) showed that CNTs are suitable for proliferation of osteosarcoma ROS 17/2.8 cells on SWCNT and MWCNT scaffolds (Zanello et al. 2006). Human osteoblast-like (Saos2) cells grown on the MWCNT scaffolds showed a higher cell density and transforming growth factor-β1 concentration compared with those grown on polystyrene and polycarbonate scaffolds (Tsuchiya et al. 2007).

Our previous studies have shown greater osteoblast differentiation on MWCNT-Ti than on Ti alone (Sirivisoot et al. 2007). Moreover, MWCNTs are a promising material for electrochemical biosensors because they also possess relatively well-characterized behavior in terms of electron transport (Padigi et al. 2007; Roy et al. 2006; Tang et al. 2004; Kurusu et al. 2006). Coating CNTs on Ti electrodes also increases the active surface area and enhances direct-electron-transfer (Liu et al. 2005, 2007). These studies encourage the use of MWCNTs to modify currently used Ti for orthopedic implants. CNTs can be produced by the laser furnace arc and chemical vapor deposition (CVD) methods. Unlike the laser furnace arc, CVD is a scalable and controllable method of obtaining a high purity of CNTs (Robertson 2004). A strong mechanical connection between CNTs and the metal is needed for nanosensor applications (Talapatra et al. 2006), and the reduced resistance between the Ti and CNTs (Ngo et al. 2004) leads to increased current passage. For example,

Sato et al. showed great potential for using CNTs in electrochemical biosensors by using CVD to grow MWCNTs with Ti bimetallic particles and cobalt (Co) as a catalyst. They revealed that Co has the ability to combine with Ti (Sato et al. 2005), and since CNTs were grown by using a cobalt catalyst in that study, a strong electrical contact between metallic Ti and MWCNTs was possible.

As mentioned, in order to form a more robust interconnection, CNTs have been anchored in the pores of anodized nanotubular Ti. MWCNTs have then been grown using CVD techniques out of the anodized Ti nanotubes (with diameters of 50–60 nm and depths of 200 nm) as a template. In vivo, many cell processes important during new bone growth rely on the redox reactions of various biomolecules and ions. The mechanisms of electron-transfer reaction and the role of proteins in aiding the electron transfer of redox processes can be examined by electrochemical analysis. In electronic theory, when two different materials come in contact with each other, electron transfer will occur in an attempt to balance Fermi levels, causing the formation of a double layer of electrical charge at the interface. Because the formation of electric contact between the redox proteins and the electrode surface is the fundamental challenge of electrochemical biosensor devices, in one study, the redox reactions of iron (II/III) and the osteoblast extracellular components at the surface of Ti, anodized Ti, and MWCNT-Ti have been investigated. How to create such novel orthopedic sensors along with the latest data concerning their use is presented below.

2 Making Orthopedic Implant Sensors

2.1 Step 1: Preparation of Nanotubular Anodized Titanium

In order to create orthopedic sensors, currently implanted titanium (Ti) has been modified to have a nanotube-structured thin layer of titanium dioxide (TiO_2) by anodization (Sirivisoot et al. 2007). Briefly, 99.2% commercially pure Ti sheets (Alfa Aesar) can be cut into squares (1 cm^2) and cleaned with acetone and 70% ethanol under sonication for 10 min each. After rinsing with deionized water, Ti can be etched for 10 s in a solution of 1.5% by weight of nitric acid and 0.5% by weight of hydrofluoric acid to remove the existing oxidized-layer on Ti. Immediately after etching, Ti can be placed as an anode electrode and a high purity platinum (Pt) mesh (Alfa Aesar) can be used as a cathode. In a Teflon beaker, both electrodes can be immersed in an electrolyte solution of 1.5% by weight of hydrofluoric acid in deionized water. The distance between Ti and the platinum mesh should be around 1 cm. A direct current power supply (3,645 A; Circuit Specialists) can then be used at 20 V between the anode and cathode for 10 min to create uniform nanotubes of TiO_2 on commercially pure Ti. It is from these uniform titanium nanotubes that electrically active sensing MWCNTs can be grown, as described next.

2.2   Step 2: Cobalt-Catalyzed Chemical Vapor Deposition for Growing MWCNTs

Afterward, MWCNTs can be grown out of the nanoporous TiO_2 by CVD. A plasma-enhanced chemical vapor deposition (PECVD) system (Applied Science & Technology Inc.) can be used to grow MWCNTs from the nanotubular anodized Ti. To do this, the anodized Ti samples can be soaked in a solution of 5% by weight of Cobaltous Nitrate (Allied Chemical) in methanol for 5 min prior to the CVD process. Then, cobalt-catalyzed anodized Ti should be rinsed with distilled water and dried with compressed air. The samples can then be placed in the CVD chamber and the air pumped out to a base pressure below 10 mTorr. The samples can be heated to 700°C in a flow of 100 standard cubic centimeters per minute (sccm) hydrogen gas (H_2) for 20 min. The gas composition can be changed to 40 sccm H_2 and 160 sccm acetylene gas (C_2H_2) for 30 min to grow MWCNTs. The MWCNT-Ti should be cooled in a 100 sccm Argon flow. The dense and entangled MWCNTs can form a three-dimensional structure on the Ti surface with exceptional electrical conductivity and surface area. In particular, conventional Ti, shown in Fig. 1a, exhibits a smooth Ti oxide surface. After anodization, nanopores of Ti oxide are formed on the Ti surface uniformly with diameters of 50–60 nm and depths of

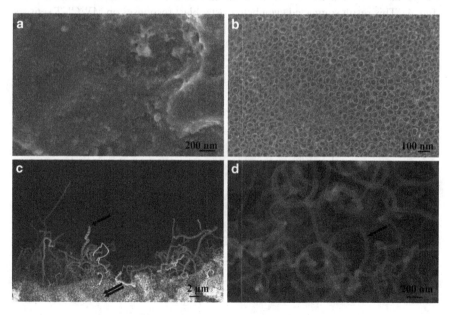

Fig. 1 SEM micrographs of the electrode surfaces: (**a**) conventional Ti; (**b**) anodized nanotu-bular Ti; and (**c**) side view and (**d**) top view of MWCNT-Ti. *Single arrow* shows MWCNTs, whereas the *double arrows* show the anodized Ti template

200 nm, as shown in Fig. 1b. After the CVD process described here, MWCNTs covered the anodized Ti, as shown in Fig. 1c, side view, and Fig. 1d, top view.

3 Biological Responses to Orthopedic Implant Sensors

Importantly, results have shown similar numbers of osteoblasts adherent after 4 h on unanodized Ti, anodized nanotubular Ti without MWCNTs, and anodized nano-tubular Ti with MWCNTs (Fig. 2); all were greater than the carbon nanopaper (used as a chemical control). This initial data was important as it indicated that the MWCNT sensor was not detrimental to osteoblast functions. Moreover, osteoblasts were observed with cytoplasmic prolongation on the surfaces of MWCNTs grown out of anodized nanotubular Ti (Fig. 3c, d), resulting in stronger adhesion (more density of the points of contact) and some might say resembling osteoblast interactions with extracellular matrix proteins (Sirivisoot et al. 2007). In contrast, osteoblasts on anodized Ti without MWCNTs were not observed to have cytoplasmic prolongation (Fig. 3a, b). Zanello et al. also observed a thin neurite-like cytoplasmic prolongation of osteoblasts that reached the nanotube bundles after cell culture for 5 days. More impressively, results have demonstrated that alkaline phosphatase activity and calcium deposition by osteoblasts increased on the MWCNTs grown from anodized nanotubular Ti sensors compared to anodized nanotubular Ti without MWCNTs, unanodized Ti, and carbon nanopaper after 21 days (Fig. 4). This result suggested even better osteoblast responses on MWCNT sensors than currently used Ti.

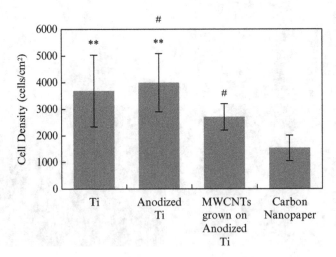

Fig. 2 Osteoblast adhesion for 4 h on Ti, Anodized Ti, MWCNTs grown on Anodized Ti, and carbon nanopaper. Values are mean ±S.E.M; $n=3$; **$p<0.05$ compared to carbon nanopaper; #$p<0.05$ compared to MWCNTs grown on Anodized Ti

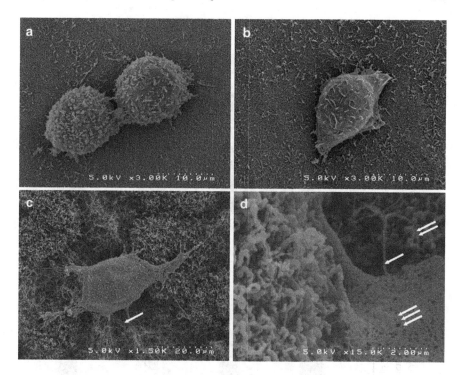

Fig. 3 Osteoblast morphology after 4 h of adhesion on: (**a, b**) Anodized Ti (scale bars = 10 μm). Round-shaped osteoblasts are on the anodized Ti substrate without MWCNTs; and (**c, d**) MWCNTs grown on anodized Ti (scale bars = 20 μm *left* and 2 μm *right*). *One-arrow* shows the cytoplasmic prolongations of osteoblasts towards MWCNTs. *Double-arrows* show MWCNTs, while *triple-arrows* show the osteoblast membrane

4 Sensing Ability of Orthopedic Implant Sensors

Of course, the next question is how well do these MWCNTs sense new bone growth? As a starting point, experiments were conducted with the $Fe(CN)_6^{4-/3-}$ redox system. The $Fe(CN)_6^{4-/3-}$ redox system with a heterogeneous one electron transfer ($n = 1$) is one of the most extensively studied redox couple in electrochemistry (Tamir et al. 2007). It has been performed on cyclic voltammetry experiments (the $Fe^{2+/3+}$ redox couple) by placing MWCNT-Ti in an electrolyte solution of 10 mM $K_3Fe(CN)_6$ and 1 M KNO_3. In potassium ferricyanide ($K_3Fe(CN)_6$), the reduction process is Fe^{3+} ($Fe(CN)_6^{3-} + e^- \rightarrow Fe(CN)_6^{4-}$) followed by the oxidation of Fe^{2+} ($Fe(CN)_6^{4-} \rightarrow Fe(CN)_6^{3-} + e^-$) under a sweeping voltage. In this study, the iron (II/III) redox couple did not exhibit any observable peaks for bare Ti or anodized Ti electrodes, as shown in Fig. 7a, b. This implied that the electrochemical reaction was slow on both of these electrodes. However, highly directed electron transfer at the MWCNT-Ti sensor electrode was observed as redox peaks, shown in Fig. 7c. At a scan rate of 100 mV/s in Fig. 7c, a well-defined redox peak appears with anodic

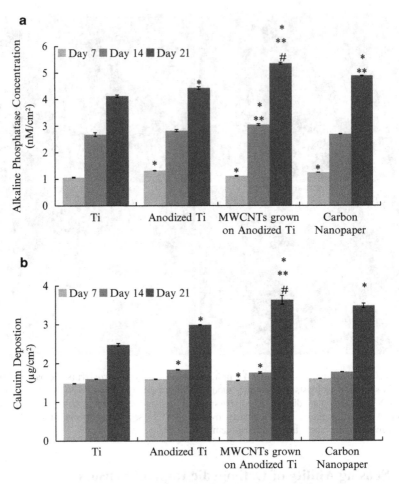

Fig. 4 Osteoblasts long-term functions: (**a**) Alkaline phosphatase activity; values are mean ± S.E.M; $n = 3$; *$p < 0.05$ compared to Ti; **$p < 0.1$ compared to anodized Ti; and #$p < 0.1$ compared to carbon nanopaper and (**b**) calcium deposition; values are mean ± S.E.M; n = 3; *$p < 0.1$ compared to Ti; **$p < 0.1$ compared to anodized Ti; and #$p < 0.05$ compared to carbon nanopaper

(E_{pa}) and cathodic (E_{pc}) potentials at 175 and 345 mV, respectively. Moreover, as shown in Fig. 7e and 9c, the relationship was linear between the anodic and cathodic peak currents versus the square root of the scan rate, while the ratio of I_{pa}/I_{pc} was about 1, corresponding to the Randles–Sevcik equation (1). Because the root scan rate had this linear relation with the peak currents, the mass transport in this process must be by diffusion. Zhang et al. found that the heterogeneous charge-transfer rate constant (k) of the $Fe(CN)_6^{4-/3-}$ complex with H_2O as a solvent was 0.05 cm/s (Zhang et al. 1991). Since the k value was in the range of 10^{-4} to 10^{-1} cm/s and $\Delta E_p > 59/n$ mV (in this case $n = 1$ and $\Delta E_p \sim 170$ mV), this process was quasireversible.

To analyze the electrochemical behavior at the surface of the MWCNT-Ti sensor electrodes, researchers have used the Randles–Sevcik equation for quasireversible processes, equation (1). Hence, the peak current (I_p) is given by:

$$I_p = 2.99 \times 10^5 A D^{1/2} n (n_a \gamma)^{1/2} C, \qquad (1)$$

where n is the number of electrons participating in the redox process, n_a is the number of electrons participating in the charge-transfer step, A is the area of the working electrode (cm^2), D is the diffusion coefficient of the molecules in the electrolyte solution (cm^2/s), C is the concentration of the probe molecule in the bulk solution (molar), and γ is the scan rate of the sweep potential (V/s). When the $Fe(CN)_6^{4-/3-}$ redox system exhibits heterogeneous one-electron transfer ($n = n_a = 1$) and the concentration C is equal to 10 mM, the diffusion coefficient D is equal to $6.7 \pm 0.02 \times 10^{-6}$ cm^2/s (Hrapovic et al. 2004; Hrapovic and Luong 2003). Hence, the quasireversible redox of iron (II/III) is truly enhanced by the novel MWCNT sensors.

To estimate how such sensors would do for bone growth, osteoblasts were cultured for 21 days and energy dispersive spectroscopy (EDS) was performed to verify the presence of the various minerals in newly formed bone. Figure 5 shows the results of one such study doing this in which the peaks of many inorganic substances, consisting of magnesium (Mg), phosphorus (P), sulfur (S), potassium (K), and calcium (Ca), were detected. The Ca/P weight ratio of minerals deposited by osteoblasts in that study on Ti (1.34) was less than that on a MWCNT-Ti sensor (1.52). However, the Ca/P ratio of hydroxyapatite (HA), the main calcium-phosphate crystallite in bone, is typically about 1.67 (Calafiori et al. 2007; Wang et al. 2003). This study demonstrated that the minerals deposited by osteoblasts on MWCNT-Ti were more similar to natural bone than the minerals deposited on Ti. X-ray diffraction (XRD) analysis also showed that more hydroxyapatite was deposited on MWCNT-Ti sensors than on both conventional and anodized Ti after 21 days of culture, as shown in Fig. 6. In addition, the amount of calcium deposited by osteoblasts as determined by a calcium quantification assay kit was 1.481 µg/cm^2 for 7 days, 1.597 µg/cm^2 for 14 days, and 2.483 µg/cm^2 for 21 days on conventional Ti, as shown in Fig. 10b. These results imply a greater deposition of calcium by osteoblasts on MWCNT-Ti sensors than currently implanted Ti.

Bone resorption and remodeling involve the secretion of hydrochloric acid (HCl) by osteoclasts (Zaidi et al. 2004). Osteoclasts dissolve bone mineral by isolating a region of the matrix and then secreting HCl and proteinases at the bone surface, resulting in the bone acting as a reservoir of Ca^{2+}, PO_4^{3-}, and OH^- minerals (Blair et al. 2002). This can be the reason that HCl is so prevalent in laboratories to dissolve calcium minerals deposited by osteoblasts on scaffolds in vitro (as the supernatant to further use in a calcium deposition assays). Thus, after dissolving calcium mineral with 0.6 N HCl, it is likely that Ca^{2+}, PO_4^{3-}, and OH^- are contained in the solution of the osteoblast extracellular components.

The formation of bone matrix minerals first depends on achieving a critical concentration of calcium and phosphorus. Then phospholipids, anionic proteins, as

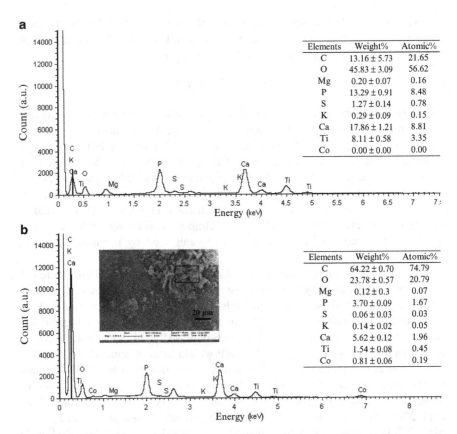

Fig. 5 EDS analysis of osteoblasts cultured for 21 days on (**a**) Ti and (**b**) MWCNT-Ti. SEM micrograph of inset (**b**) shows the analyzed area. For the MWCNT-Ti, more Ca and P deposited by osteoblasts were observed. Tables in (**a**) and (**b**) show the composition of the mineral deposits after osteoblasts were cultured for 21 days. The Ca/P weight ratio on bare Ti was 1.32 and on MWCNT-Ti was 1.52

well as calcium and phosphorus aggregate in nucleation pores that are in the 35 nm "hole-zone" between collagen molecules (Bronner and Farach-Carson 2003). The addition of calcium, phosphate, and hydroxyl ions contributes to the growth of crystalline hydroxyapatite. However, the crystals are not pure since they also contain carbonate, sodium, potassium, citrate, and traces of other elements, such as strontium and lead. The imperfection of hydroxyapatite contributes to its minerals' solubility, which plays a role in the ability of osteoclasts to resorb the mineral phases. Although osteoblasts synthesize type I collagen, which is the predominant organic components of bone, type III/V/VI collagen also exist in bone. Moreover, noncollagenous proteins in bone (such as growth factors, osteocalcin, osteopontin, and osteonectin) and proteins in serum and other tissues (such as fibronectin, vitronectin, and laminin) are absorbed into the mineral component during bone growth. These components are directly involved in the genesis and maintenance of bone. Thus, not only minerals, but also other noncollagenous proteins, are broken

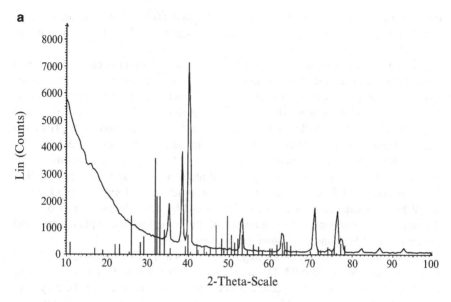

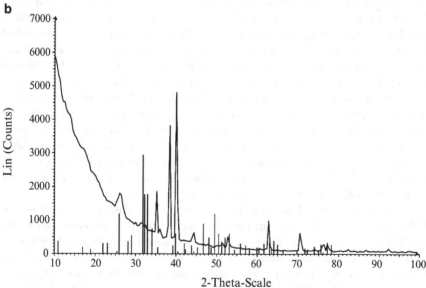

Fig. 6 XRD analysis of hydroxyapatite-like (HA; $Ca_5(PO_4)_3OH$) deposited minerals after osteoblasts were cultured for 21 days on (**a**) Ti and (**b**) MWCNT-Ti. The micrographs show that the peak pattern of HA more closely matches that of the mineral deposited by osteoblasts when cultured on MWCNT-Ti than Ti

down after dissolution with HCl. As such, nonspecific proteins may exist in the supernatant of bone formation when cyclic voltammetry is performed on MWCNT-Ti sensors and may contribute to the observed redox process shown in this chapter. On an electrode surface, the native conformation of a protein may be retained (reversible or diffusion-controlled) or distorted (irreversible or adsorption-controlled), depending on the extent of the interactions (such as the electrostatic or covalent bonds) between them.

The type of electrolyte is important to the redox reaction because the gained capacitance and scan rate window are dependent on it (Hong et al. 2002). Studies to date have not degassed with nitrogen or argon when performing cyclic voltammetry with MWCNT sensors, thus, the redox reactions in CV are related to the H_2O decomposition, which includes H^+ reduction and OH^- oxidation. The pH of the electrolyte also affects the H_2O decomposition. In the case of an acidic 1 N HCl solution (without other electroactive molecules), the H^+ ions induced the formation of H_2 gas at -0.2 V as a result of the increased reduction current (Hong et al. 2002). In the alkaline 1 M KOH solution, OH^- ions resulted in the formation of O_2 gas at 0.7 V by the oxidation process. This fact should be noted because in vivo, these decompositions may occur at the interface of the Ti implant due to pH changes and the presence of H_2O and O_2 around the implant.

Other studies have shown promising results when using CNT-modified electrodes in biosensing. Since MWCNTs have good electrochemical characteristics as electron mediators and adsorption matrices (Sotiropoulou et al. 2003), they may further enhance applications in biosensor systems. For example, Harrison et al. suggested that CNTs offer a promising method to enhance detection sensitivity because they have high signal-to-noise ratios (Harrison and Atala 2007). The structure-dependent metallic character of CNTs should allow them to promote electron-transfer reactions for redox processes, which can provide the foundation for unique biochemical sensing systems at low overpotentials (Roy et al. 2006). The electrolyte-electrode interface barriers are reduced by CNTs because they facilitate double-layer-effects (Fang et al. 2006). Typically, when the supporting electrolyte is at least a 100-fold greater than the active electrolyte (Christian 1980), the charge in the electrolyte solution causes the Debye layer to be more compact. Therefore, this compact layer can rapidly exchange electrons between electroactive proteins and the surface of an MWCNT-Ti electrode. This is likely the reason the sharpened cathodic and anodic peaks in CV have been observed, as shown in Figs. 7c, 8c, 9b.

Although the area of the working electrode is constant in the presence of MWCNTs, the overall surface area and the electroactivity of the surface increases. The MWCNTs act as a nanobarrier between the titanium oxide layer, which was its growth template, and the electrolyte solution. The surface atoms or molecules of MWCNTs play an important role in determining its bulk properties due to their nanosize effects (Cao 2004). A large surface, corresponding to the greater electrocatalytic activity, confers the catalytic role on the MWCNTs in the chemical reaction. Well-defined and persistent redox peaks are shown in Fig. 7c, confirming the increases in electron transfers and higher electrochemical activities at the MWCNT-Ti surface. Cyclic voltammetry was also performed with a glassy carbon electrode (GCE; Bioanalytical) and a platinum electrode (PTE; Bioanalytical), and it also showed $\Delta E_p > 59/n$ mV (data not shown). Importantly, for the same scan rate, the CV of GCE and PTE showed redox reactions with similar but more widely separated anodic and cathodic peaks, confirming the performance of the MWCNT-Ti electrode.

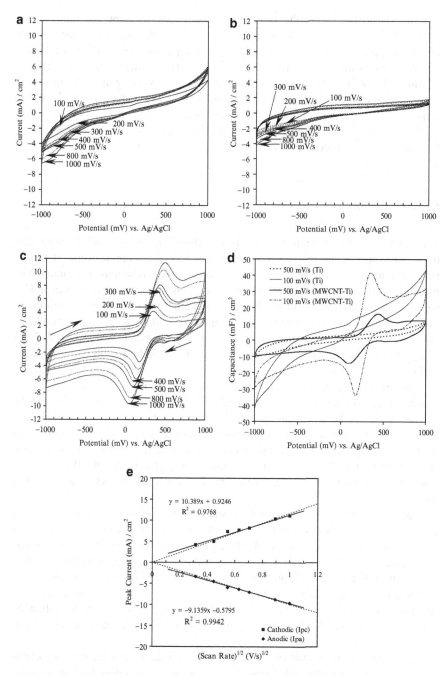

Fig. 7 Cyclic voltammograms with an electrolyte solution of 10 mM $K_3Fe(CN)_6$ in 1 M KNO_3 for: (**a**) conventional Ti; (**b**) anodized Ti; and (**c**) MWCNT-Ti. (**d**) The capacitance of all the electrodes in comparison. (**e**) The plot of the square root of scan rates with anodic peak currents (I_{pa}) and cathodic peak currents (I_{pc})

However, in Fig. 7a, b, the Ti and anodized Ti did not show any redox peaks. This is likely due to the inhibiting property of the titanium oxide on electron transfer. Figure 7a, b shows that bare Ti has less capacitance than MWCNT-Ti. The capacitance relationship is derived from: $i = C(\mathrm{d}v / \mathrm{d}t)$, where $\mathrm{d}v / \mathrm{d}t$ is a scan rate (Tamir et al. 2007). Hence, the capacitance between the working and reference electrodes during cyclic voltammetry was calculated by dividing the current in CV with respect to the specific scan rate, as shown in Figs. 7d and 9d. In summary, the electrochemical response of the MWCNT-Ti electrode promoted a higher charge transfer than the Ti and anodized Ti electrodes.

5 Discussion

The results from the previous section showed the utility of MWCNT-Ti as an electrode for detecting $Fe^{2+/3+}$ redox couples. For biological applications, it is also necessary to show that MWCNT-Ti can detect redox reactions associated with osteoblast differentiation. The extracellular components from osteoblasts in an HCl solution have been used to mimic the biological environment around an orthopedic implant. Figure 8a, b show that the bare Ti and anodized Ti electrodes cannot detect any redox process. In contrast, the results in Fig. 8c confirmed that MWCNTs enhanced the direct electron transfer through Ti by adding a highly conductive surface the MWCNTs.

After decreasing the surface area of the MWCNT-Ti electrode from 1 cm² to 1 mm², the redox potentials also decreased as shown in Figs. 8c and 9b. The faradic current of the oxidation process dropped approximately ten times with respect to the decrease in surface area A, corresponding to equation (1). Corresponding with the Randles–Sevcik equation, the peak current was linearly proportional to the area. When plotting the anodic (I_{pa}) and cathodic peak (I_{pc}) currents of MWCNT-Ti, a linear relationship to the square root of the scan rates was observed, as shown in Fig. 9c.

The CV showed well-defined redox peaks in the osteoblast supernatants at all concentrations. The peaks detected in the solution for the calcium concentrations of 1.481 µg/cm² (after 7 days), and 1.597 µg/cm² (after 14 days) have less faradic current and I-V graph area than the concentration of 2.48 µg/cm² (after 21 days), as shown in Fig. 10a, b. Typically, the more HA deposited and synthesized by osteoblasts, the higher the calcium and protein concentrations. Therefore, it is likely that the supernatant from 21 days had higher protein concentrations than the solutions from 7 and 14 days, leading to the stronger redox reactions and higher capacitance between the electrode surfaces.

In addition, interpretation of CV results must consider other factors. In particular, a measurement of the redox reactions when oxygen was dissolved in the electrolyte solution showed that the peak currents were shifted toward the negative. Furthermore, a change in the pH of the electrolyte solution and the presence of

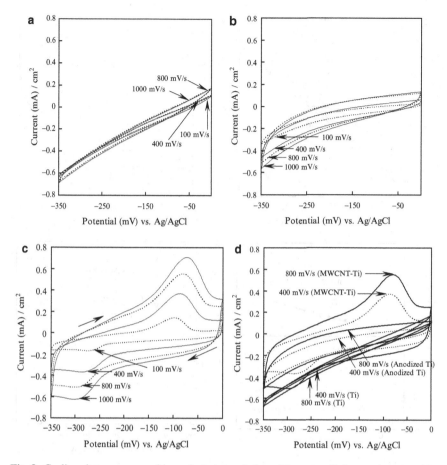

Fig. 8 Cyclic voltammograms with an electrolyte solution of the extracellular matrix secreted by osteoblasts after 21 days of culture for: (**a**) conventional Ti; (**b**) anodized Ti; and (**c**) MWCNT-Ti with a working area of 1 cm². (**d**) CV of all three electrodes in comparison. Only MWCNT-Ti possessed the quasireversible redox potential, while conventional Ti and anodized Ti did not

water may also shift the current and affect the potential of the redox peaks. Thus, CV with the potential range between −1 and 1 V on MWCNT-Ti was performed in a pure solution of 0.6 N HCl without other proteins or ions. An H⁺ reduction peak at a potential between −0.5 and −0.8 V was found (data not shown). Moreover, the H⁺ reduction peak still appeared when window ranges from −5 to 5 V and from −0.35 to 0 V were used, confirming the H_2O decomposition occurred in our CV experiments.

It is clear that the MWCNT-Ti electrodes showed a potential for better performance than commercially available electrodes, such as GCE and PTE. When using the sweep voltage from −1 to 1 V, a set of mixed redox peaks with two oxidations and two reductions appeared (data not shown). These two redox peaks in the CV, indicated

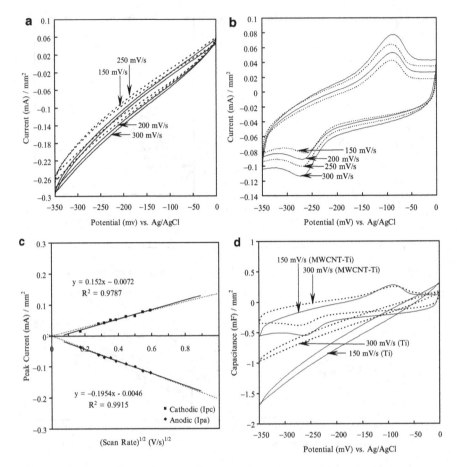

Fig. 9 Cyclic voltammograms with an electrolyte solution of the extracellular matrix secreted by osteoblasts after 21 days of culture for (**a**) conventional Ti and (**b**) MWCNT-Ti with a working area of 1 mm². (**c**) Plot of the experimental cathodic and anodic peak currents, obtained from (**b**), versus the square root of the scan rates; and (**d**) the compared capacitance of MWCNT-Ti and Ti

that the process likely involved more than one electron transfer. For each couple, the amplitude and position of the oxidation peak in CV was unequal to those of the reduction peak. The second oxidation peak (omitted from Figs. 8c or 9b) was observed at 236 mV for the MWCNT-Ti electrode, 405 mV for the GCE, and −33 mV for the PTE, all with the scan rate of 100 mV/s. However, for the GCE and the PTE electrodes, a second reduction did not appear, while the MWCNT-Ti showed two complete and well-defined redox reactions. These two peaks reflected that more than one electron transfer was involved, since one occured at a positive potential and one at a negative in the CV. Further, the current amplitude for the MWCNT-Ti peaks was noticeably larger than that for GCE and PTE. Thus, it can be concluded that the MWCNT-Ti electrode has higher electron transfer as well.

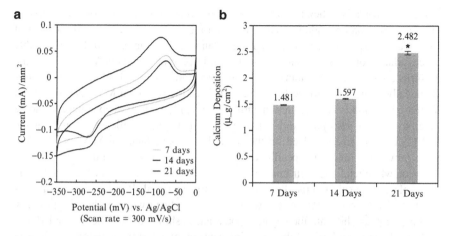

Fig. 10 (**a**) Cyclic voltammograms of MWCNT-Ti electrodes in an electrolyte solution of the extracellular matrix secreted by osteoblasts cultured on conventional Ti after 21 days. (**b**) Results from the calcium deposition assay determined the calcium concentrations in an electrolyte solution of the extracellular matrix secreted by osteoblasts on conventional Ti after 7, 14, and 21 days of culture. Data = mean ± S.E.M; $n = 3$; *$p < 0.01$ (compared to 7 and 14 days)

At first it can be hypothesized that these redox reactions happened due to the redox of inorganics (such as Ca or P). In the case of Ca, the reduction of Ca^{2+} to Ca crystals (nucleation) typically is expected with the reduction current for a highly negative potential. A positive feedback IR compensation (IRC) needs to be applied to the CV in order to drive the oxidation current to dissolve the Ca crystal into Ca^{2+}. For example, an overpotential of more than −1 V, IRC of 0.771 Ω, melting temperature of 900°C, and scan rate of 50 mV/S were required (Chen and Fray 2002). Without applying IRC in this study, the diffusion-controlled behavior, which electrochemically depletes calcium cations near the electrode surface, has been impossible to achieve. In addition, the reduction and oxidation of titanium oxides, which are the reaction counterparts of calcium redox, appeared at a positive potential between 0 and 0.25 V, respectively, which did not appeared in our CV studies. In the case of P, the cyclic voltammetry experiments were performed further in phosphate buffered saline (PBS). A final concentration of 1 M PBS, which was prepared in-house, had 137 mM NaCl, 10 mM phosphate, 2.7 mM KCl, and a pH of 7.4. The metal orthophosphate ions exhibited a high stability to reduce from P(V) (such as PO_4^{3-}, PO_4^-) to P(III), or further to P(I). However, no redox peaks appeared in the potential range of −1 to 1 V (data not shown). It is therefore likely that neither the redox reactions of calcium ions nor phosphate ions occurred due to the absence of the peaks. Instead, it has been hypothesized that the appearance of the redox peaks when using an MWCNT-Ti electrode is likely because of the reduction and oxidation of proteins from the osteoblast supernatant.

The observed quasireversible redox reaction reflected the fact that MWCNT-Ti sensors promoted electrochemical reactions. However, an irreversible process can occur at the interface of an electrode due to the denaturation of proteins, or ion reduction. Typically, when the carbon bonds of proteins are altered, only irreversible organic reduction or oxidation happens (Schuring 2000). While the configuration of ions (Ca^{2+}, PO_4^{3-}, OH^-), which remain unchanged, also can lead to the irreversible process due to only its reduction or oxidation. Importantly, without MWCNTs on the Ti surface, nonspecific protein adsorption may readily occur, leading to an electrochemically insulating surface layer. Nevertheless, the redox on MWCNT-Ti did not decrease its oxidation and reduction current after applying the sweep voltage for some time, so protein adsorption did not occur apparently in significant amounts at the MWCNT-Ti electrode. Indeed, in the osteoblast supernatant, the MWCNT-Ti electrode provides a relatively specific surface to which proteins can bind reversibly by diffusion while retaining their function. The protein redox is also dependent on the pH of the electrolyte solution (acidic or basic) (Battistuzzi et al. 1999). This shift appears due to the influence of 0.6 N HCl with H_2O as a solvent which influences proteins, leading the peak shift into the negative potential, as shown in Figs. 8c, 9b, and 10a.

It has been hypothesized that proteins were involved in these redox reactions at the surface of MWCNT-Ti in the solution of osteoblast extracelluar components. If those proteins maintain their function and are detected by the MWCNT-Ti electrode, the excellent electrochemical behavior explained here may show that MWCNT-Ti sensors are superior candidates for biosensor applications. However, investigating the type and size of acidic resistant proteins is still in progress.

As shown here, the electronic performance of MWCNT-Ti sensors, but not the bare Ti, provides excellent redox reactions. As mentioned, the electrochemical experiments were performed so far without degassing with bubble nitrogen gas. This omission may have resulted in an apparent background signal, which was mixed and appeared in current-voltage curves (CV). Moreover, the redox reaction at the MWCNT-Ti sensor mentioned here may depend on various ion sources, including reversible protein adsorption, dissolved oxygen, and the acidic nature of HCl. However, the redox reactions of iron (II/III) and the osteoblast extracellular components still occurred only when using MWCNT-Ti sensors, and not when using Ti or anodized Ti. Importantly, it was shown that MWCNT sensors performed as the electrocatalysts for the oxidation and reduction of iron (II/III) and of extracellular components, which included various proteins and inorganic substances synthesized and deposited by osteoblasts.

The electrolyte solution of osteoblast extracellular components was composed of inorganic and organic substances. This solution was not classified in terms of the types of molecules in solution, but rather, all components were dissolved together in the acidic solution and used in all cyclic voltammetry experiments in this study. Future work needs to classify the type of proteins in such osteoblast supernatants.

However, it is likely that in vivo the capacitance and the electroactive surface of Ti will be enhanced after bone tissue is formed without the presence of MWCNTs. After hip implant insertion, blood and body fluids surround the Ti implant, creating a specific capacitance for the Ti electrode in the beginning. Within 1 month, as osteoblasts deposit calcium around the implant, the capacitance of the implant and bone tissue increases. Cell membranes also have electric potentials and perform redox process (Jeansonne et al. 1978, 1979), and it has been found that after 14 days in vitro these redox processes at the surface of bare Ti increased.

In vitro, nevertheless, MWCNTs have been used to modify the electrodes to detect other biomolecules and enhance their redox reactions, such as NADH (Lin et al. 2005), hydrogen peroxide (Kurusu et al. 2006; Lin et al. 2005), glucose (Liu et al. 2005), putrescine (Luong et al. 2005), and DNA (He et al. 2006). Specifically, the growth of MWCNTs from nanoporous metals electrodes can enhance many diverse sensor applications. For example, MWCNTs functionalized with thiol groups have been used for sensing aliphatic hydrocarbons (such as methanol and ethanol), forming unique electrical identifiers (Padigi et al. 2007). Moreover, MWCNTs grown on silicon substrates enhanced ionic conductivity and have been integrated with unmodified plant cellulose as a film, and used as a cathode electrode in a lithium ion battery, or as a supercapacitor with bioelectrolytes (such as biological fluid and blood) (Pushparaj et al. 2007). Interestingly, both supercapacitors and batteries derived from a MWCNTs-nanocomposite film, can be integrated to build a hybrid, or dual-storage battery-in-supercapacitor device. The cytocompatible MWCNTs-device may be useful to integrate with an orthopedic-implant biosensor as passive energy storage.

Not only can an implanted electronic circuit be powered by an implant battery (fabricated or self-integrated) within the implant material, but it also can be induced by an external power supply. For orthopedic applications, telemetric devices started in 1966 and became imperative as a telemetric orthopedic implant in order to transmit a signal to an external device (Kaufman et al. 1996; Bergmann et al. 2001a, b; D'Lima et al. 2005; Graichen and Bergmann 1991; Graichen et al. 1999). For example, Graichen et al. (2007) developed a new 9-channel telemetry transmitter used for in vivo load measurements in three patients with shoulder Ti endoprosthesis. The low radio frequency (RF) of an external inductive power source generates power transmission through the implanted metal. At the end of the Ti implant, the pacemaker feedthrough rounds a single loop antenna to transmit the pulse interval signal, which is modulated at a higher RF by the microcontroller, to an external device. Then, the RF receiver of the external device synchronizes with the modulated pulse interval to recover the data stream and report to a clinician. Thus, either an implanted battery or a telemetry system can supply the electrical power to a calcium-detectable chip inside the orthopedic implant for clinical use.

The automatically-sensing concept for bone growth juxtaposed to orthopedic implants can also be applied for determining infection and scar tissue formation. An energy source energizes a telemetric-implanted circuit, and allows it to transmit data to an external receiver to determine bone growth. Indeed, the electrochemical biosensor may reduce the complexity of imaging techniques and patient difficulties.

6   Conclusions

The capabilities of using titanium (Ti) as an electrochemical electrode have been increased remarkably by growing MWCNTs out of anodized Ti nanopores (MWCNT-Ti). MWCNTs improved the sensitivity of the bare Ti electrode displaying redox peaks in cyclic voltammograms and interestingly high capacitance. Such results provide evidence that MWCNT-Ti can serve as a novel electrochemical electrode with exceptional electrocatalytic properties due to increased surface area and conductivity. Moreover, MWCNTs were shown to be more electroactive in chemical transformations than the metallic (Ti) surface, which typically undergoes electrochemical oxidation or dissolution of metal oxides, but is chemically susceptible to corrosion. MWCNTs promoted a redox reaction by enhancing the direct electron transfer through their electrically conductive surfaces surrounded by ionic solutions, which contained the electroactive species, herein ferri/ferricyanide and the extracellular components from osteoblasts. Moreover, MWCNTs are cytocompatible, promoted osteoblast differentiation after 21 days, and can be integrated into a supercapacitor or battery to enhance the functionalities of biosensing systems in vivo. Therefore, MWCNT-Ti sensors are an exceptional candidate as an electrochemical electrode to determine in situ new bone growth surrounding an orthopedic implant. Further, the ability of MWCNT-Ti to sense osteoblast extracellular components by detecting their redox reaction profiles in specific concentrations may improve the diagnosis of orthopedic implant success or failure, and thus improve clinical efficacy.

Acknowledgments The authors would like to thank the Coulter Foundation for funding some of the results in this chapter.

References

Aoki N, Yokoyama A, Nodasaka Y, Akasaka T, Uo M, Sato Y, et al. Cell Culture on a Carbon Nanotube Scaffold. Journal of Biomedical Nanotechnology 2005;1:402–405(404).

Balani K, Anderson R, Laha T, Andara M, Tercero J, Crumpler E, et al. Plasma-sprayed carbon nanotube reinforced hydroxyapatite coatings and their interaction with human osteoblasts in vitro. Biomaterials 2007;28(4):618–624.

Battistuzzi G, Loschi L, Borsari M, Sola M. Effects of nonspecific ion-protein interactions on the redox chemistry of cytochrome c. Journal of Biological Inorganic Chemistry 1999;4(5):601–607.

Bergmann G, Deuretzbacher G, Heller M, Graichen F, Rohlmann A, Strauss J, et al. Hip contact forces and gait patterns from routine activities. Journal of Biomechanics 2001;34(7):859–871.

Bergmann G, Graichen F, Rohlmann A, Verdonschot N, van Lenthe GH. Frictional heating of total hip implants. Part 1: measurements in patients. Journal of Biomechanics 2001;34(4):421–428.

Blair HC, Zaidi M, Schlesinger PH. Mechanisms balancing skeletal matrix synthesis and degradation. Biochemical Journal 2002;364(2):329–341.

Bronner F, Farach-Carson MC. Bone formation. New York: Springer, 2003.

Calafiori A, Di Marco G, Martino G, Marotta M. Preparation and characterization of calcium phosphate biomaterials. Journal of Materials Science: Materials in Medicine 2007;18:2331–2338.

Cao G. Nanostructures & nanomaterials: synthesis, properties & applications. London: Imperial College Press, 2004.

Chen GZ, Fray DJ. Voltammetric studies of the oxygen-titanium binary system in molten calcium chloride. Journal of the Electrochemical Society 2002;149(11):E455–E467.

Chen Y, Zhang TH, Gan CH, Yu G. Wear studies of hydroxyapatite composite coating reinforced by carbon nanotubes. Carbon 2007;45(5):998–1004.

Christian GD. Analytical chemistry. New York: Wiley, 1980.

Ciombor DM, Aaron RK. The role of electrical stimulation in bone repair. Orthobiologics 2005;10(4):579–r593.

D'Lima DD, Townsend CP, Arms SW, Morris BA, Colwell CW. An implantable telemetry device to measure intra-articular tibial forces. Journal of Biomechanics 2005;38(2):299–304.

Fang WC, Sun CL, Huang JH, Chen LC, Chyan O, Chen KH, et al. Enhanced electrochemical properties of arrayed CNx nanotubes directly grown on Ti-buffered silicon substrates. Electrochemical and Solid State Letters 2006;9(3):A175–A178.

Giannona S, Firkowska I, Rojas-Chapana J, Giersig M. Vertically aligned carbon nanotubes as cytocompatible material for enhanced adhesion and proliferation of osteoblast-like cells. Journal of Nanoscience and Nanotechnology 2007;7:1679–1683(1675).

Graichen F, Bergmann G. Four-channel telemetry system for in vivo measurement of hip joint forces. Journal of Biomedical Engineering 1991;13(5):370–374.

Graichen F, Bergmann G, Rohlmann A. Hip endoprosthesis for in vivo measurement of joint force and temperature. Journal of Biomechanics 1999;32(10):1113–1117.

Graichen F, Arnold R, Rohlmann A, Bergmann G. Implantable 9-channel telemetry system for in vivo load measurements with orthopedic implants. IEEE Transactions on Bio-Medical Engineering 2007;54(2):253–261.

Harrison BS, Atala A. Carbon nanotube applications for tissue engineering. Cellular and Molecular Biology Techniques for Biomaterials Evaluation 2007;28(2):344–353.

He P, Xu Y, Fang Y. Applications of carbon nanotubes in electrochemical DNA biosensors. Microchimica Acta 2006;152(3):175–186.

Hong MS, Lee SH, Kim SW. Use of KCl aqueous electrolyte for 2 V manganese oxide/activated carbon hybrid capacitor. Electrochemical and Solid-State Letters 2002;5(10):A227–A230.

Hrapovic S, Luong JHT. Picoamperometric detection of glucose at ultrasmall platinum-based biosensors: preparation and characterization. Analytical Chemistry 2003;75(14):3308–3315.

Hrapovic S, Liu Y, Male KB, Luong JHT. Electrochemical biosensing platforms using platinum nanoparticles and carbon nanotubes. Analytical Chemistry 2004;76(4):1083–1088.

Iijima S. Helical microtubules of graphitic carbon. Nature 1991;354(6348):56–58.

Jeansonne BG, Feagin FF, Shoemaker RL, Rehm WS. Transmembrane potentials of osteoblasts. Journal of Dental Research 1978;57(2):361–364.

Jeansonne BG, Feagin FF, McMinn RW, Shoemaker RL, Rehm WS. Cell-to-cell communication of osteoblasts. Journal of Dental Research 1979;58(4):1415–1423.

Kaufman K, Kovacevic N, Irby S, Colwell C. Instrumented implant for measuring tibiofemoral forces. Journal of Biomechanics 1996;29(5):667–671.

Kurusu F, Koide S, Karube I, Gotoh M. Electrocatalytic activity of bamboo-structured carbon nanotubes paste electrode toward hydrogen peroxide. Analytical Letters 2006;39:903–911.

Lin Y, Taylor S, Li H, Fernando KAS, Qu L, Wang W, et al. Advances toward bioapplications of carbon nanotubes. Journal of Materials Chemistry 2004;14(4):527–541.

Lin Y, Yantasee W, Wang J. Carbon nanotubes (CNTs) for the development of electrochemical biosensors. Frontiers in Bioscience : A Journal and Virtual Library 2005;10:492–505.

Liu Y, Wang M, Zhao F, Xu Z, Dong S. The direct electron transfer of glucose oxidase and glucose biosensor based on carbon nanotubes/chitosan matrix. Biosensors and Bioelectronics 2005;21(6):984–988.

Liu S, Lin B, Yang X, Zhang Q. Carbon-nanotube-enhanced direct electron-transfer reactivity of hemoglobin immobilized on polyurethane elastomer film. Journal of Physical Chemistry B 2007;111(5):1182–1188.

Luong JHT, Hrapovic S, Wang D. Multiwall carbon nanotube (MWCNT) based electrochemical biosensors for mediatorless detection of putrescine. Electroanalysis 2005;17(1):47–53.

Ngo Q, Petranovic D, Krishnan S, Cassell AM, Ye Q, Li J, et al. Electron transport through metal-multiwall carbon nanotube interfaces. IEEE Transactions on Nanotechnology 2004;3:311–317.

Padigi SK, Reddy RKK, Prasad S. Carbon nanotube based aliphatic hydrocarbon sensor. Biosensors and Bioelectronics 2007;22(6):829–837.

Pushparaj V, Shaijumon M, Kumar A, Murugesan S, Ci L, Vajtai R, et al. Flexible energy storage devices based on nanocomposite paper. Proceedings of the National Academy of Sciences of the United States of America 2007;104(34):13574–13577.

Robertson J. Realistic applications of CNTs. Materials Today 2004;7(10):46–52.

Roy S, Vedala H, Choi W. Vertically aligned carbon nanotube probes for monitoring blood cholesterol. Nanotechnology 2006;17(4):S14–S18.

Sato S, Kawabata A, Kondo D, Nihei M, Awano Y. Carbon nanotube growth from titanium-cobalt bimetallic particles as a catalyst. Chemical Physics Letters 2005;402(1–3):149–154.

Schuring J. Redox fundamentals, processes, and application. New York: Springer, 2000.

Sirivisoot S, Yao C, Xiao X, Sheldon BW, Webster TJ. Greater osteoblast functions on multi-walled carbon nanotubes grown from anodized nanotubular titanium for orthopedic applications. Nanotechnology 2007;18(36):365102.

Smart SK, Cassady AI, Lu GQ, Martin DJ. The biocompatibility of carbon nanotubes. Toxicology of Carbon Nanomaterials 2006;44(6):1034–1047.

Sotiropoulou S, Gavalas V, Vamvakaki V, Chaniotakis NA. Novel carbon materials in biosensor systems. Biosensors and Bioelectronics 2003;18(2–3):211–215.

Supronowicz PR, Ajayan PM, Ullmann KR, Arulanandam BP, Metzger DW, Bizios R. Novel current-conducting composite substrates for exposing osteoblasts to alternating current stimulation. Journal of Biomedical Materials Research 2002;59(3):499–506.

Talapatra S, Kar S, Pal SK, Vajtai R, Ci L, Victor P, et al. Direct growth of aligned carbon nanotubes on bulk metals. Nature Nanotechnology 2006;1:112–116.

Tamir G, Moti B-D, Itshak K, Raya S, Ze'ev RA, Eshel B-J, et al. Electro-chemical and biological properties of carbon nanotube based multi-electrode arrays. Nanotechnology 2007(3):035201.

Tang H, Chen JH, Huang ZP, Wang DZ, Ren ZF, Nie LH, et al. High dispersion and electrocatalytic properties of platinum on well-aligned carbon nanotube arrays. Carbon 2004;42(1):191–197.

Tsuchiya N, Sato Y, Aoki N, Yokoyama A, Watari F, Motomiya K, et al. Evaluation of multi-walled carbon nanotube scaffolds for osteoblast growth. In: Tohji K, Tsuchiya N, Jeyadevan B, editors. AIP, New York; 2007:166–169.

Wang H, Lee J-K, Moursi A, Lannutti JJ. Ca/P ratio effects on the degradation of hydroxyapatite in vitro. Journal of Biomedical Materials Research Part A 2003;67A(2):599–608.

Webster TJ, Waid MC, McKenzie JL, Price RL, Ejiofor JU. Nano-biotechnology: carbon nanofibres as improved neural and orthopaedic implants. Nanotechnology 2004;15(1):48–54.

Wei W, Sethuraman A, Jin C, Monteiro-Riviere NA, Narayan RJ. Biological Properties of Carbon Nanotubes. Journal of Nanoscience and Nanotechnology 2007;7:1284–1297(1214).

Zaidi M, Moonga BS, Huang CL-H. Calcium sensing and cell signaling processes in the local regulation of osteoclastic bone resorption. Biological Reviews 2004;79(01):79–100.

Zanello LP. Electrical properties of osteoblasts cultured on carbon nanotubes. Micro & Nano Letters 2006;1(1):19–22.

Zanello LP, Zhao B, Hu H, Haddon RC. Bone cell proliferation on carbon nanotubes. Nano Letters 2006;6(3):562–567.

Zhan C, Kaczmarek R, Loyo-Berrios N, Sangl J, Bright RA. Incidence and short-term outcomes of primary and revision hip replacement in the United States. Journal of Bone and Joint Surgery. American Volume 2007;89(3):526–533.

Zhang Y, Baer CD, Camaioni-Neto C, O'Brien P, Sweigart DA. Steady-state voltammetry with microelectrodes: determination of heterogeneous charge transfer rate constants for metalloporphyrin complexes. Inorganic Chemistry 1991;30(8):1682–1685.

Index

T.J. Webster (ed.), *Nanotechnology Enabled In situ Sensors for Monitoring Health*,
DOI 10.1007/978-1-4419-7291-0, © Springer Science+Business Media, LLC 2011